"In this insightful and enjoyable book, Michelle Janning draws readers into a deeper understanding of love letters as cultural artifacts. Utilizing innovative methods, this timely contribution tells an illuminating story about the ways in which we curate love letters—as reflections of collective values and individual experiences—that will appeal to a wide range of readers."

—Adina Nack, *Professor, Sociology, California Lutheran University*

"Janning has produced a fascinating book exploring the cultural practice of writing, sending, and saving love letters—even in an age when pen and paper have given way to emails and text messages. It is a theoretically engaged, yet accessible, reminder that while expressions of love may take different forms throughout time, they remain an indelible part of our personal lives and romantic experiences."

—David J. Hutson, *Assistant Professor, Sociology, Pennsylvania State University, Abington*

"*Love Letters* offers a lively read full of fascinating insights, not only about how we think about romantic relationships but also about preservation, place, and nostalgia in everyday life. Janning weaves these insights into a new framework for understanding culture and connection in a time of rapid technological change. And she proves an expert guide to understanding the broader social context shaping our most personal stories."

—Lyn Spillman, *Professor, Sociology, University of Notre Dame*

"If we ever thought, as many of us did, that the digital age would crush romantic messages or make them evanescent, Janning has shown us that we were wrong; love just surfaces in a different form. This is a wonderful book full of rich and surprising details and very suitable for classes. Students can learn a lot about the way culture works by reading about love from diverse positions in the social structure, by differences in sexual orientation, by differences in romantic experience and life histories, and finally through the different gendered perspectives."

—Pepper Schwartz, *Professor, Sociology, University of Washington*

LOVE LETTERS

In today's world of Tinder and texting, do we write and save love letters anymore? Are we more likely to save a screenshot of a text exchange or a box of paper letters from a lover? How might these different ways to store a love letter make us feel? Sociologist Michelle Janning's *Love Letters: Saving Romance in the Digital Age* offers a new twist on the study of love letters: what people do with them and whether digital or paper format matters. Through stories, a rich review of past research, and her own survey findings, Janning uncovers whether and how people from different groups (including gender and age) approach their love letter "curatorial practices" in an era when digitization of communication is nearly ubiquitous. She investigates the importance of space and time, showing how our connection to the material world and our attraction to nostalgia matter in actions as seemingly small and private as saving, storing, stumbling upon, or even burning a love letter. Janning provides a framework for understanding why someone may prefer digital or paper love letters, and what that preference says about a person's access and attachment to powerful cultural values such as individualization, patience in a hectic world, longevity, privacy, and preservation of cherished things in a safe place. Ultimately, Janning contends, the cultural values that tell us how romantic love should be defined are more powerful than the format our love letters take.

Michelle Janning received her Ph.D. in Sociology from the University of Notre Dame. She is Professor of Sociology at Whitman College in Walla Walla, Washington. She has published numerous book chapters and articles on family relations and material culture, authored the book *The Stuff of Family Life: How Our Homes Reflect Our Lives* (2017), and edited the collection *Contemporary Parenting and Parenthood: From News Headlines to New Research* (2018). She has received a Fulbright Specialist Grant and teaching awards, and her work has appeared in national and international television, radio, Internet, and print outlets, including *U.S. News and World Report*, *Real Simple*, *The Verge*, *Author Story*, and Positive Parenting Radio. Go to www.michellejanning.com to learn more.

Routledge Series for Creative Teaching and Learning in Anthropology
Editor: Richard H. Robbins, SUNY Plattsburgh and
Luis A. Vivanco, University of Vermont

This series is dedicated to innovative, unconventional ways to connect undergraduate students and their lived concerns about our social world to the power of social science ideas and evidence. We seek to publish titles that use anthropology to help students understand how they benefit from exposing their own lives and activities to the power of anthropological thought and analysis. Our goal is to help spark social science imaginations and, in doing so, open new avenues for meaningful thought and action.

Books in this series pose questions and problems that speak to the complexities and dynamism of modern life, connecting cutting edge research in exciting and relevant topical areas with creative pedagogy.

Available

Love Letters
Saving Romance in the Digital Age
Michelle Janning

The Baseball Glove
History, Material, Meaning, and Value
David Jenemann

Persian Carpets
The Nation as a Transnational
Commodity
Minoo Moallem

An Anthropology of Money
A Critical Introduction
Tim Di Muzio and Richard H. Robbins

Coffee Culture, 2e
Local Experiences, Global Connections
Catherine M. Tucker

Forthcoming

Seafood
From Ocean to Plate
Richard Wilk & Shingo Hamada

LOVE LETTERS

Saving Romance in the Digital Age

Michelle Janning

Routledge
Taylor & Francis Group

NEW YORK AND LONDON

First published 2018
by Routledge
711 Third Avenue, New York, NY 10017

and by Routledge
2 Park Square, Milton Park, Abingdon, Oxon, OX14 4RN

Routledge is an imprint of the Taylor & Francis Group, an informa business

Library of Congress Cataloging-in-Publication Data
A catalog record for this book has been requested

ISBN: 978-1-138-05525-4 (hbk)
ISBN: 978-1-138-05526-1 (pbk)
ISBN: 978-1-315-16600-1 (ebk)

Typeset in New Baskerville
by Apex CoVantage, LLC

For Christina and Richard, who model the best parts of love,
and who brought the love of my life into this world

CONTENTS

FIGURES

SERIES FOREWORD

The premise of these short books on the *Anthropology of Stuff* is that stuff talks, that written into the biographies of everyday items of our lives—coffee, T-shirts, computers, iPods, flowers, drugs, and so forth—are the stories that make us who we are and that make the world the way it is. From their beginnings, each item bears the signature of the people who extracted, manufactured, picked, caught, assembled, packaged, delivered, purchased, and disposed of it. And in our modern market-driven societies, our lives are dominated by the pursuit of stuff.

Examining stuff is also an excellent way to teach and learn about what is exciting and insightful about anthropological and sociological ways of knowing. Students, as with virtually all of us, can relate to stuff, while at the same time discovering through these books that it can provide new and fascinating ways of looking at the world.

Stuff, or commodities and things, are central, of course, to all societies, to one extent or another. Whether it is yams, necklaces, horses, cattle, or shells, the acquisition, accumulation, and exchange of things is central to the identities and relationships that tie people together and drive their behavior. But never, before now, has the craving for stuff reached the level it has; and never before have so many people been trying to convince each other that acquiring more stuff is what they most want to do. As a consequence, the creation, consumption, and disposal of stuff now threaten the planet itself. Yet to stop or even slow down the manufacture and accumulation of stuff would threaten the viability of our economy, on which our society is built.

This raises various questions. For example, what impact does the compulsion to acquire stuff have on our economic, social, and political well-being, as well as on our environment? How do we come to believe that there are certain things that we must have? How do we come to value some commodities or form of commodities above others? How have we managed to create commodity chains that link peasant farmers in Colombia or gold miners in Angola to wealthy residents of New York or teenagers in Nebraska? Who comes up with the ideas for stuff and how do they translate those ideas into things for people to buy? Why do we sometimes consume stuff that is not very good for us? These short books examine such questions, and more.

PREFACE

In this book, I investigate paper and digital love letters as cultural objects. As you continue to read, you'll see that the study of love letters as artifacts that tell stories of human relationships is alive and well across many academic disciplines, from history to literature, from sociology to anthropology, and from communication studies to philosophy. I am not alone in my aim to include what people communicate to each other, and how and why they do it, in the body of research used by scholars to uncover how human social life actually operates. A lot of this past research has focused on historical understandings (and misunderstandings) of love letters, often centering on the content of letters and how social conventions have dictated how the letters were supposed to be written (e.g., with formality or informality, with content focusing on different topics, with norms about amount of time between letters, or with the goal of sharing privately or with others). My aim is to combine elements and ideas from scholars of a variety of academic fields along with my own research in order to investigate contemporary love letters in a new way. Importantly, while past research informs my interpretations, I do not examine the content of love letters, nor do I investigate archived collections of letters from decades long ago in order to delve into historical differences that shape content and writing conventions. I take a bit of a different path, one that branches into explorations of love letter format and everyday practices used to manage the letters.

I investigate *what people do* with their love letters, with a focus on two topics: (1) the meaning of everyday "curatorial" practices (tasks similar to those used by someone organizing an art collection) used to manage saved love letters; and (2) the impact of digital and paper love letter format. By examining people's curatorial practices surrounding love letters once they have them—the saving, storing, displaying, organizing, and sometimes dispossessing of them—I hope to give voice to the letter-writers and letter-keepers, since my research asks people to talk about letters that are collected and managed by the possessors themselves rather than catalogued in an archive or collection managed by someone else. By focusing on love letter format (digital or paper), I hope to add to our understanding of the impact of technology on contemporary human relationships and on the cultural definition of romantic love.

I also aim to continue my larger intellectual project of studying household possessions and their use in order to uncover how identities, roles, relationships, and group differences and inequalities can be shown in our material world, especially in light of massive technological change. I've done this with mothers' and fathers' roles in managing physical and digital photo albums (Janning and Scalise 2015), kids' connections to their bedrooms and technology when their parents get divorced (Janning, Collins, and Kamm 2011), calendars and other objects that traverse work and home boundaries (Janning 2009), and college students' use of objects to differentiate childhood and adulthood identities (Janning and Volk 2017). These past studies, as well as abbreviated investigations into love letters that have been published in shorter and more technical formats (Janning and Christopherson 2015; Janning 2015; Janning, Gao, and Snyder 2017), form the foundation for this book.

In the pages that follow I take you on a journey into the everyday curatorial practices of saving, storing, and revisiting paper and digital love letters as symbols of romantic relationships. I discuss the influence of information and communication technologies (ICT) on love letter saving practices, with an eye toward uncovering how these practices may differ within and between demographic groups based on gender, race, nationality, age, socioeconomic status, and sexuality. Through the elaboration of my U.S.-based quantitative and qualitative survey research findings, as well as via anecdotes, news stories, and overviews of others' research, the book delves into the everyday practices and habits that govern how we think of intimate relationships and love. I focus on the symbolic representation of romance via material cultural objects and spaces, as well as on the way nostalgia and time are socially constructed in an era where digital communication has redefined interpersonal relationships, romance, the blurry boundary between public and private lives, and the preservation of individual and collective memory.

ACKNOWLEDGMENTS

In a way, this section should be read as my love letter to everyone who has made this project possible to think about, craft, edit, and promote.

I'm most grateful to my husband Neal, with whom I can share intimate inside jokes in a text exchange about our past handwritten love letters. And our son Aaron, who so far has only received a couple notes from admirers passed to him in school, continues to push me to ask good questions, do good research, and write good books. He also puts up with my dumb jokes and tells a fair share of his own, which keeps me smiling. Both of these patient people have endured countless dinnertime conversations where I say, "Okay I'm done talking about my book. But wait, I have another question . . ."

The Whitman College community, including my wonderful classroom students, research assistants, and faculty and staff colleagues, continues to provide me with the resources and inspiration to do the work of researching and writing. In particular, I'm grateful for my sociology colleagues Gilbert Mireles, Alissa Cordner, Helen Kim, Keith Farrington, Bill Bogard, and Alvaro Santana-Acuña, especially since my writing projects took me away from the classroom and departmental service for two semesters. I'm grateful to Sarah Hurlburt for our discussions about the perforated boundaries between literary analysis, history, sociology, and material culture studies. The college generously provided me with a sabbatical and research funding so that I had time to write, and so that I was able to involve undergraduate students at all stages of my scholarship. Thanks especially to Emma Snyder, Ailie Kerr, and Wenjun Gao, whose assistance was invaluable in the creation of this project, the collection of data, the organization of ideas, and the writing. Thank you also to the handful of sociology students who came over to my house for pizza during a semester that I wasn't teaching in order to listen to me present my findings and conceptual framework, offer suggestions, ask questions, and assure me that the ideas presented in this book would help students learn how to think about their social worlds.

To the people who have shared survey responses and informal stories about your love letters (or your grandpa's love letters, or your ex-boyfriend's letters, or the letters of famous people that you think are interesting), thank you. This book is an unfinished project simply because I will never tire of hearing stories about love letters written, saved, stored, lost, found, burned, and even organized alphabetically or meticulously by date. Your stories started my curiosity, and they keep fueling it.

I'm grateful to people who have read and offered feedback on the manuscript, especially Jennifer Steffens, Yvonne Janning, and Erik Zimmerman (and thanks to all of the Routledge and *Anthropology of Stuff* series colleagues). You all have tolerated my attempts to be clear and thorough with patience, good ideas for edits, and kind words to keep me going.

For all of my friends and family—it is the love stories that we have shared and continue to create that personalize this project for me, whether it's fondly reminiscing about the past or carefully treading on difficult terrain to make sure we're all doing okay today. Whether we write carefully crafted notes or send group texts to keep in touch, I love you all.

And to Maggie the dog: thank you for lying next to my writing desk and twitching your paws as you dreamed about running through fields. You were a miracle. What a gift that you lived long enough to help me write a book about love. I miss you and your loving eyebrows.

1

THE STUFF OF LOVE

The Historical and Cultural Significance of (Saving) Love Letters

It would be difficult to find someone in today's world who has never heard or seen a reference to love letters on a big or small screen, in print, or on the radio. I am part of a social network filled with people who know about love letters as they have been represented across time and in various media forms, from plays to contemporary podcasts, from 18th-century epistolary novels to romantic comedies and love songs. It was no surprise, then, that when I solicited ideas for fictional or non-fictional references to love letters in my social media feed, I got more responses than I would ever have time to read, view, listen to, or skim. Suggestions such as *The Notebook, You've Got Mail,* and *The Lakehouse* were offered to show love letters' (both paper and digital) prominence on the big screen (and sometimes in print first).

Books and plays, such as Shakespeare's *Romeo and Juliet,* A. S. Byatt's *Possession: A Romance,* or Pierre Choderlos de Laclos's *Les Liaisons Dangereuses,* were listed in order to show how romance—and the letters that connote it—have featured centrally in authors' writings for hundreds of years (and, actually, for some of these, on the big screen, too). Scholar friends reminded me that key social thinkers often referenced in academic courses have said some of their most important things in letters, evidenced by writings from Simone de Beauvoir, Charlotte Perkins Gilman, and Max Weber, among others. Even books that have emerged from love letter references in other platforms are being published, as evidenced by the two-volume series *Love Letters of Great Men;* this is a published collection of real letters from prominent historic male figures such as Beethoven and Lord Byron that was spawned from a reference in the fictional movie *Sex and the City.* It wasn't a real book until the movie mentioned it.

My musician friends suggested songs such as "Love Letters," the Academy Award–nominated song featured in the 1945 movie of the same name. This song has been so popular that its lyrics—"I'm not alone in the night when I can have all the love you write"—have been crooned over the years by artists such as Elvis Presley, Nat King Cole, and Diana Krall. Another friend called my attention to podcasts in order to show that love letters (and the saving of them) are being discussed within the milieu of current hot topics.

One such podcast—Helen Zaltzman's *The Allusionist*—aired a two-part series on letters in October 2017. In this duet of podcasts, Zaltzman shares stories of writing and saving letters from lovers, family members, and even strangers. She begins the first episode with her own story:

> I had a relationship before the 21st century. Neither of us had the internet yet, or mobile phones, or even landlines sometimes. So we wrote letters to each other, a lot, for more than three years. And I still have them, in a box, in another box. The odd thing is, I only have half the story, his half; he's got mine. Or maybe he doesn't, maybe he did get rid of them, a long time ago; I really hope he did destroy them. I haven't looked at his letters since we broke up half my lifetime ago; but I can't bring myself to throw them away either . . . I, however, very rarely write letters now. I barely even send birthday or Christmas cards. But I can't shake the idea that letters are important.

As the podcast continues, Zaltzman interviews others whose stories shed light on how love letters, and physical hold-in-your-hand letters generally, matter, even if they're not letters between lovers. For example, she shares a story about a letter-writing program designed to give letters from listeners to San Quentin Penitentiary inmates whose stories are featured on the podcast *Ear Hustle*. This program is meant to humanize the inmates and connect them to the world outside of the prison, a task that is especially challenging given their limited access to cell phones and internet. Nigel Poor, the program's coordinator, reveals the significance of the program:

> Getting a letter from the outside world is like a treasure, you know. It has the power that mail had originally. . . [when] it took months to get a letter, and when you got a letter you just savored it. It was a really important object, and that's how it is inside prison. If you get a letter it's like gold.

Zaltzman also shares the story of Dave Nedelberg, whose accidental stumbling across a love letter he never sent in high school (back in 1991) was the impetus for his stage show (and now multi-media platform) *Mortified*, where people read things they wrote as kids. He reflects on the impact that finding that box in his parents' home in Michigan has had on his life today:

> And inside that box, AAAAHH! It was really like opening up the Ark of the Covenant. Anyway there was like a whole bunch of things in there, and one of those things was a letter. And I read it and I thought, "This

is ridiculous." . . . It's a pretty seismic change this little tiny letter had in my life. Whatever my 16-year-old brain thought going on a date with her would lead to, I got something a whole lot better. It never ceases to amaze me where this letter has taken me. Because of this letter I've met you. Because of this letter I've traveled the world. I've met like huge heads of various industries. I've met people who said that they've cried and had their lives changed because of something that happened in *Mortified* that they experienced. Babies have been born because of *Mortified*. And all these extraordinary things. That letter had no idea how powerful it was. Because it's just a stupid little thing saying "this is who I am."

These stories demonstrate a few key themes that are woven throughout this book. First, as Zaltzman's own story and the San Quentin letter-writing program show, technology matters in any research about love letters, since the nearly ubiquitous availability of digital communication (except in some places, such as prisons) has affected the frequency and format of love letter writing, sharing, and storing; it has also changed the very meaning of a paper letter as something that may be even more treasured given its rarity. The role of technology in how love letters are defined is the focus of Chapter 2. What people do with their letters once received (whether they keep them, where they store them, and how often they look at them) may reveal what the letters mean to their recipients, which is the focus of Chapter 3. But these "curatorial" practices—as with Dave's storage of his childhood letters and other mementos in his parents' home—may also tell the larger story of cultural norms surrounding saving and storing material and digital objects, as well as the importance of spaces and places in an understanding of people's relationships, life stages, identities, and the changing definition of romantic love. As Dave's story shows, how people think about themselves today is wrapped up in these letters, especially when they reminisce about the past and wonder about the relevance of sentimentality and nostalgia in current identities. This—the cultural construction of time and memory and the impact of nostalgia—is featured in Chapter 4. Society is filled with group differences that can translate into inequalities that manifest in seemingly unimportant mundane practices such as love letter writing, practices that tell fascinating individual stories but that also tell the larger story about how culture works. This connection between the personal world of romantic communication and the collective world of shared cultural *values* (beliefs held about what is important or expected in a group), including the way that privilege shapes both individual and group experiences, is the focus of Chapter 5.

Through the examples listed above it is apparent that today's cultural milieu is filled with reference to the significance of written communication between people who love each other. But has this changed over time? How might the changing definition of love, both across time and among different groups, matter in the definition of love letter today? How do sociologists and anthropologists approach their research on love as an emotion and as a part of culture? And, since countless scholars and writers have examined the content of love letters extensively in past work, what can focusing on love letters as objects of cultural significance, which includes what people do with them, add to current understandings?

Importantly, my aim here and in every chapter is not to assert whether saving love letters is right or wrong, how best to organize and "curate" them if they are saved, or whether digital or paper love letters are better or worse to save. My aim is to uncover what people do, why they do it, and what it may say about larger cultural forces that shape behaviors and beliefs.

But first, I define love letter.

What Is a Love Letter?

Pen-and-paper messages sent across geographic distance between lovers are the go-to image of love letters in popular media, advice columns about how to have a good relationship, and conversations I've had with people as I've ventured into my love letters research (see Figure 1.1). There's even a card game entitled Love Letters that contains this kind of imagery. In this game, the players act as competing suitors to a princess who is locked in a palace. Each player woos the

Figure 1.1 **Pen-and-Paper Letter** (Source: Author Photograph)

princess by getting their letters to her before the others. The imagery on the cards includes roses, calligraphy on parchment sealed with a red wax stamp, and cleavage-revealing clothing from an era where princesses are presumed to have hung out in towers. The cards come in a red velvet bag with the words "Love Letter" embroidered in gold cursive across the bag. The images in these game pieces call to mind ink pens and cursive on special paper, the color red, sexuality, and the function of a love letter to make romantic feelings known to the recipient. Nowhere in the instructions does it define love letter or explain its function or importance in the larger culture. Nowhere in the instructions is there an explanation for the imagery used in the game's illustrations. It's simply a game of luck and strategy that implements imagery of a subject that is assumed to be familiar among those who play the game.

Later in this chapter I delve into definitions of romantic love, including how the definition must be understood in terms of how culture works. Romance, intimacy, and partner connectedness have become central to the thematic content of contemporary love letters and their associated imagery in games and other popular culture elements, but these themes must be historically and geographically situated. And, as with any object of communication, what people are expected to do, what they have the ability to do, and what motivates them to do things are all impacted by forces beyond individual minds, bodies, and desires. For now, and even though it seems as if I'm defining something without first explaining the complex ways in which the concepts that are intimately comingled with it have changed and been challenged, I offer a working definition of love letter to be used throughout the rest of the book.

What exactly is the *object* in a book on love letters? How inclusive should a definition of love letter be, both so that it has enough focus to offer a sense of the concept's boundaries, and enough flexibility to account for the list of seemingly endless platforms for communication in today's rapidly changing digital society? Should it include envelopes, ink, or internal phone hardware used in digital texting apps? Certainly one may consider a smartphone itself to be a love letter-as-object if it contains romantic messages in one of its platforms, but a phone (or computer or box or sock drawer or cloud storage service) is a *container* that holds the love letter, not the letter itself. The hardware and applications in a phone, like ink and paper, are tools used to craft, send, store, and access the message, not the message itself. Additionally, certainly photographs, audio recordings, intricate drawings or doodles, and even recorded video messages or exchanges could be considered love letters, but my focus in this book is on exchanges that involve *written* communication.

I include only written messages as a way to represent what Stanley (2015) refers to as "ordinary writing, [which] has been a continuing feature in many cultures" (241). This means that I consider a love letter to be *words written down and exchanged with someone else where the intent of the written message is, at least in part, to convey romantic love, intimacy, affection, or sexual interest.* The love letter can be one word, or it can be a typed multi-page rhapsody. It can be handwritten or typed (or, I suppose, could even contain glued-on letters cut out of a magazine). It can contain other elements such as drawings or photos. I consider the message itself to be the love letter, in whatever format it comes. Included in the definition of love letter above is the notion that a letter can take different forms, have different lengths, include different norms about both style and content, include content unrelated to the conveyance of romantic love, and necessitate different time frames for receipt and response. While it is not the case that "anything goes" in this definition, it is true that it is expansive and allows for a lot of variety and interpretation. Thus, both digital and physical forms of messages that convey romance and affection count in my definition, which, in the survey findings that I present throughout this book, includes paper letters, notes, and cards, as well as text messages, emails, Facebook messages, and even captured Snapchats.

As you will read later in the book, this definition is not universally agreed upon, with the disagreement depending on someone's belief that certain elements must be present in order to consider it to be a love letter. Further, since I conducted my survey in 2013, a myriad of other digital platforms that are used to exchange intimate messages between couples (or prospective sex or dating partners) have emerged and/or gained popularity, including Tinder, Couplete, LoveByte, Twyxt, and Kouply, to name a few. These are platforms couples use to communicate, and many have features whereby their messages can be saved and organized. Some platforms, such as Tinder (a social networking application used to signify to geographically close fellow users whether you find them intriguing enough to meet up in real life or not, indicated by swiping right or left on a smartphone), even have suggestions for how to communicate within the app in order to find a successful match with a person, suggesting that message conventions are alive and well in letter writing today. In the rest of the book, then, it is important to keep these and other new platforms in mind as I present findings. In many ways I suspect that the types of communication platforms I include in my research are similar to these newer platforms in terms of the themes I investigate, but it is important to consider how future research could examine differences within new digital platforms in terms of these themes.

My definition aligns with how love letters have been depicted in media forms and scholarly research for the last century, often invoking imagery of men at war sending love notes home to women, noted figures engaged in secret love affairs using only the written word to convey their feelings for each other, and the occasional Shakespeare sonnet embedded among other words of devotion that may be spritzed with a little perfume on purchased stationery. Popular love letter imagery has also included inked drawings of hearts and acronyms for phrases signifying physical attraction or contact such as SWAK (Sealed With A Kiss) or, as Teo (2005: 347) found in his study of Australian love letters, EGYPT (Eager to Grasp Your Pretty Toes), INDIA (Intercourse Needed Day I Arrive), or 0410E (Oh, For One Naughty). Now, of course, the convention of using pen and paper is less common, and the imagery doesn't just evoke heterosexual relationships. And by virtue of digital communication technology, the usual expectation of awaiting an expected response for days or weeks rather than seconds or minutes has been abandoned. But the content devoted to attraction, intimacy, and affection remains in academic and popular renditions of (and popular sentiment about) love letters. I maintain that this content—that of romantic love—is included in a definition of love letter.

What Is Cultural About Romantic Love?

Romantic love has varied in its definition across time and geographic space. The ideal ways to both experience and express love are culturally variable. In the Western world, for example, the industrial period "romantic love ethic" idealized falling head over heels in love and losing control. In places such as India, as Arlie Hochschild describes in *The Commercialization of Intimate Life* (2003), romantic love is more likely to be "seen as a dangerous, chaotic emotion" (122) because it destabilizes systems that require marriages to be arranged by extended families, often with economic considerations at the forefront of the arrangement. The feeling of dread and guilt among soon-to-be spouses that researchers studying romantic love in places that practice arranged marriages have studied alters the feeling of romantic love itself. Thus, there are cultural differences in how romantic love operates in terms of how present it is in modern relationships, and in terms of how it actually feels.

As Hochschild continues, "[e]ach culture has its unique emotional dictionary, which defines what is and isn't, and its emotional bible, which defines *should* and *shouldn't*" (122). Necessarily, then, it is important to look at romantic love as an emotion that is experienced and expressed culturally. In the cultural context of people whose stories and experiences are discussed in this book—that is, contemporary adults in the United States between the ages of 18 and 86—there has been an increased focus on romantic love as emotionally fulfilling and expressive, and yet, simultaneously precarious. People are told to aspire to a "richly communicative, intimate, playful, sexually fulfilling love. But, at the same time, the social context itself warns against trusting such a love too much"

(123). In this sense, the culture of romantic love prescribes values such as communicativeness and the expression of feelings at the same time it prescribes some fear that it may be lost.

So, culture matters. But what exactly is culture? Here I ground my research on love letters as *cultural objects* and as symbols that indicate *how culture works* in society. I rely extensively on the definition of culture helpfully elaborated by sociologist Ann Swidler (2001) in her book *Talk of Love: How Culture Matters*.

The phrases *in my culture* or *what culture are you from* have made their way into many conversations between people trying to get to know each other, and this use of the term mostly closely resembles how I discussed contemporary Western or Indian culture in the previous paragraphs. This is a way of thinking about culture that connotes someone's connection to a larger group by virtue of their similar lives. The similarity may be based on language, food, racial-ethnic identity, nationality, religious practices, ways of establishing kinship ties, geographic location, or lifestyle. Classic anthropological writings by Clifford Geertz and others have captured this definition as "the entire way of life of a people, including their technology and material artifacts, or . . . everything one would know to become a functioning member of a society" (Swidler 2001: 12). The use of the term *culture* in this way shows up in book titles and everyday conversation in forms such as *American culture* or *skateboarding culture*. This use separates out a collection of people that, for better or worse, have been identified (by others or by themselves) as having something important in common. In other words, to use culture in this way is to suggest it is a group status or entire way of life that exists outside of an individual person, and that it doesn't change much.

Beck and Beck-Gernsheim (2014) offer the helpful reminder in their study of families whose lives are geographically separated across the globe that, while it is tempting to use the term culture to refer to people from a particular place that is often indicated by a common ethnicity or language or nationality (or practice of arranged marriage), the existence of couples and families whose intimate lives transcend national borders and group boundaries lessen the accuracy of this definition. For example, to say that there is such a thing as rural culture or French culture (or French rural culture) is handy in casual conversation, but it is not good enough at capturing variation within groups and places. Thus culture, rather than being defined entirely as a static or fixed way of grouping people together, is better defined as dynamic.

The use of the term *culture* to refer to a whole way of life for a group of people has been amended by those who favor a dynamic definition as

> the publicly available symbolic forms through which people experience and express meaning. . . [Symbolic forms are the means through which

learning and sharing happen, and include phenomena such as] beliefs, ritual practices, art forms, . . . ceremonies, . . . language, gossip, stories, and rituals of daily life.

(Swidler 2001: 12)

This definition depicts culture as more fluid and allows for more individual *agency* (the ability of a person to have autonomy and control in life, including performing actions that shape and shift culture); at the same time it acknowledges that there are shared public symbols, and even a cultural system, that can convey and transmit meaning collectively.

More specifically related to how individuals play out romantic love as a part of culture in their everyday lives, Swidler points out that, "from a sociologist's point of view, love is a perfect place to study culture in action" (2). She suggests that it is in people's *use* of culture where the mechanisms of how culture operates at both collective and individual levels can be seen, or how "culture actually becomes meaningful for people—when they are put to use and in what ways" (12). Similarly, Spillman (2001) articulates the importance of understanding culture by investigating people's meaning-making processes, often uncovered by examining their everyday practices. In this sense, people "do culture." Only by studying the varied ways that people enact, use, practice, and experience symbols—how someone "participates in a particular culture" (Swidler 2001: 13)—will a complete understanding of how culture works be made possible. In terms of love letters, then, understanding how culture works requires knowing about societal norms, rules, and values about intimate relationships and written communication, but it also requires knowing how people enact, refute, or amend those norms, rules, and values in their everyday practices. To study whether and how people save, store, organize, revisit, or throw away love letters as cultural symbols can help describe how the culture of contemporary romantic love works.

So, how do I define romantic love? While hundreds of writers over centuries have written about the elusive concept of romantic love[1] I utilize Teo's (2005) definition: "the wide range of expressions and practices of affectionate attraction between . . . people. . . [distinguished] from other types of love such as love between family members or friends" (343). In order to understand how romantic love operates in terms of culture, I turn to a topic that, at first, seems highly centered on the individual rather than on larger entities such as culture, but that actually gets at both: *emotion*.

Love, romantic or not, is defined in many places as an emotion. Emotions are studied by those interested in the inner-workings of people's hearts and minds (psychologists come to mind, as do those interested in the biological and chemical changes in bodies when people experience romantic love). For this book, I'm more interested in how people's definition of love is shaped by forces outside of their individual minds and bodies, forces that are squarely situated in the realm of culture. In particular, I look at how romantic love, as an emotion, is cultural. I use Hochschild's (2003) definition of emotion: "the awareness of bodily cooperation with an idea, thought, or attitude and the label attached to that awareness" (75). I also presume, like most anthropologists and sociologists, that ideas, thoughts, attitudes, and labels do not just exist at the individual level. They are embedded in how culture works. The following excerpt from Mesquita, Boiger, and De Leersnyder (2016: 34) highlights the role that emotion plays in the dynamic definition of culture:

> Emotions are iterative and active constructions that help an individual achieve the central goals and tasks in a given (cultural) context. Adopting a perspective of action means that the research naturally shifts to the ways cultures afford and constrain how people "do" emotions, and away from culture as a one-time socializing force that shapes the emotions people "have." Culture, then, becomes a framework within which people jointly and collectively do emotions in interactions and collectives, people construct those emotions that help them achieve "collective intentionality."

While emotions and the expressions of emotions seem like very personal experiences, they actually represent how people enact culture in their lives. Emotions are both individual and collective. And, since the expression of romantic love (and even, as Hochschild explains, the way that love *feels*) has varied across time and geographic space, it is clear that emotions, like culture, are simultaneously dynamic expressions of agency and collections of expectations that come from larger social forces. To illustrate, take the example of using a term of endearment, discussed in the 2013 BBC News article "Languages of Love: 10 Unusual Terms of Endearment." If culture didn't matter at the collective level, the symbolic expression of romantic love could either be expressed differently by every single person on the planet (thus making the expression entirely individualized), or could be expressed exactly the same in France, where *Mon Petit Chou* ("My Little Cabbage") is a term of endearment, as it is in the United States, where "Honey" is not uncommon (thus making the expression universal). While reference to food-based terms is present in both of the above linguistic demonstrations of love to a partner, the range of food options to reference in the term of endearment is prescribed by a collective and shared culture. But, importantly, there is room for individual agency even in this scenario, since it is up to the individual to determine whether she calls her lover a cabbage or not.

As exemplified by the use of a term of endearment, one of the ways that people "do emotions" is to use material and linguistic symbols to communicate those emotions to an audience. Love letters operate both as pieces of material culture and as means of communication that offer people ways to express emotion to each other individually. The collective part comes into play when things such as the marketplace (pressure for people to buy roses on Valentine's Day to show romantic love in a culturally acceptable way), media (movies that show happy endings when characters adhere to cultural expectations about the expression of romantic love through thoughtful communication [e.g., *You've Got Mail*]), and even politics (news stories about politicians who send messages to lovers outside of marriage and/or that may contain illicit sexual images of their own bodies) are considered. Communicating romantic love to someone via a love letter, thus, is both an individual and a collective action.

Connecting Love and Letters

Scholars interested in uncovering cultural norms and values of a particular time period and place have often looked to letter archives to do their digging. This is referred to as *epistolary* (relating to letters) research and is still used today in contemporary ethnographic, historical, and linguistic research (e.g., Tamboukou 2011).[2]

Judith Coffin (2010), in her review of the thousands of letters written by Simone de Beauvoir (feminist author of the classic *The Second Sex*) in the middle of the 20th century, notes that it was the postwar transformation in Europe that provided a backdrop for letter writers in that region to lose some inhibitions in their ability to broach sensitive topics (such as marriage troubles and unwanted pregnancies and children) and intimate desires (including those that relate to homosexuality). Writing letters by hand also connoted more formality than typing them at the time of de Beauvoir's writing. Lest one assumes that multi-tasking is limited to today's households, Coffin also tells a story about a letter that included a shopping list on the back for a party that de Beauvoir hosted for a few friends—a list that included champagne, vodka, whiskey, foie gras, and caviar. Grocery items on one side of a piece of paper, words of affection on the other.

Coffin asserts that de Beauvoir's letters

constitute a remarkable archive of interior lives during the 1950s, testifying to the persistence of a nineteenth-century regime of pain, confusion, and ignorance in the sexual lives of many, to the fresh blasts of air of a changing cultural climate, and to the efforts—some stammering, some clear and unabashed—to speak about sexual feeling—or in more modern terms, to acknowledge desire.

(1065)

These letters, then, "capture a turning point in the history of sexual discourse and feeling" (1067). The layers of significance are plenty here. First, there is a historical epistolary record of how romance, love, intimacy, and sexuality are being played out in a particular temporal and cultural milieu. Second, the record is itself evidence of how written communication has served to be the conduit of cultural change, especially concerning gender roles and sexuality. In other words, the content of love letters tells how people think and talk about love in a particular time and place; the letters themselves are objects of cultural significance that are the locations for the expression of love.

What goes into a love letter? To some extent, this is a question of content. Like books, letters are usually *about* some sort of content: the weather, someone's health, daily goings-on, and expression of feelings about any of these. In that case, a love letter needs to be categorized as pertaining to romantic love. A response to the question about what constitutes a love letter also depends heavily on people's understanding of the term *love*, which the previous section detailed as another historically and geographically situated concept.

To another extent, what goes into a letter also calls to mind how it's written—its formality, aesthetic appearance, and grammatical style, for instance. There are norms about what's contained in the letters called the "epistolary pact" (Lyons 1999: 233). This set of conventions includes items such as "specific forms of address and farewell which determine the tone of the relationship, encouraging familiarity or establishing distance" (Lyons 1999: 236), as well as "expectations about the frequency of correspondence, length of the letter or the number of pages, handwriting, and the sacrifice made to maintain regular and substantial correspondence" (Teo 2005: 347).

The conventional epistolary elements of a love letter depend, of course, on what's defined as normal in a particular time and place in terms of values, appropriate expressions of an emotion such as romantic love, and preferred writing styles. The pen-and-paper letter that many people think of when I talk to them about love letters—handwritten, on paper, adhering to formal written syntax conventions, sent with a precise address, and noted with a day/month/year time stamp—is a historically and geographically specific object that, as with letters in general, gained popularity in the 18th and 19th centuries in wealthy northern countries. It is a cultural object with a particular connection to changes in industry and technology (including the U.S. Postal Service and pen and paper production companies, which today compete with wireless and cellular service providers and email platforms). However, the goal of the love letter—to communicate romantic love in proxy form to someone who is not present in order to get a response—has been around for millennia. This matters because when researchers ask people whether they think certain formats of communication should count as love letters, their response depends on whether their definition focuses more on the stylistic conventions or the goals of the letter as evidenced in its content.

In many ways, this book is an example of epistolary research, though, as I describe throughout, I focus less on content and writing conventions, and more on format and practices associated with managing one's love letters, which I refer to as "curatorial practices."

Love Letter Curatorial Practices

How romantic love is experienced as an emotion at both the individual and collective levels can be revealed by examining what people do with a love letter once received. Love letter curatorial practices are as important to study as love letter content and conventions, though the focus of epistolary and archival research has tended to avoid studying these elements because the letters are often not situated in their original storage locations (if they're saved), and because it is impossible to ask the original writers and receivers from decades or centuries ago about their current everyday practices surrounding household objects such as letters.[3] Curatorial practices—how and where people save, store, re-read, organize, display, and perhaps choose to throw away—represent the main focus of my research.

When I talk with people about my love letters research they delve into anecdotes that usually fall into one of three categories: (1) they have a precise spot where love letters are stored and can go into detail about the storage spot or receptacle (sometimes with an offer to show me the letters, the storage spot, or both); (2) they no longer have some or all of their love letters, with varying degrees of control over their loss, from intentional burning to unplanned tossing them out with sadness after a disastrous basement flood; or (3) they have no idea where their love letters are, and they express feeling proud, guilty, or anxious as a result.

Objects such as love letters have their own stories, sometimes referred to as *cultural biographies* (Belk 2013). Possessions as *extended selves* (objects that reflect people's identities) serve as identity and memory markers. Intentionally saving objects gives them a kind of sacredness, which stems from the high level of emotional involvement imbued in the cultural biography of the object (Kopytoff 1986) and the significance of the object to the saver's identity. Hepper, Ritchie, Sedikides, and Wildschut (2012) articulate that saved objects help people remember past experiences (both one-time experiences and transitions throughout life stages), relationships, and selves, which in turn helps them define their current selves. Perhaps most relevant for a project on love letters, saved objects and keepsakes can trigger nostalgic emotional responses connected to those experiences and changes, in either positive or negative ways. An object becomes a trigger when it accumulates a history from the social interactions in which it is involved. When looking at a past love letter, the accumulated history embodied in it can invoke the recalling of a version of the self as salient in the past relationship, as well as stir emotions about that relationship that may impact how someone feels today (Hepper et al. 2012).

People's behaviors affect how and whether objects become part of their extended selves. Belk (2013) calls the acts of organizing, curating, preserving, using, and perhaps dispossessing saved objects "rituals of possession and disposition" (479–480). He argues that people's actions as they engage with an object matter in determining the significance of the object as salient for identities. Put another way, what people do with their stuff demonstrates what it means to them and what it says about their past and present identities. Thus, examining curatorial practices with objects heightens understanding of relationships, roles, and values. What people do with love letters shows how identities, relationships, and roles work in today's society.

Conclusion

With this book I hope to round out current understandings about the content and conventions of love letter writing by adding format and curatorial practices in the mix. The changing impacts of technology, along with the ever-evolving everyday practices that demonstrate values, are featured centrally in the questions guiding this book, the answers I offer throughout, and the questions that remain.

Culture is dynamic, produced by people in their everyday lives both individually and collectively. People participate in the production of culture when they organize family photos in an album to show to their (current or future) grandchildren. They select which pictures to include and exclude based on what messages they convey about everything from how much emotion to show in pictures to who gets included in a picture labeled "family." These are not curatorial decisions made in a vacuum; including some pictures and excluding others are impacted by cultural values. People also participate in cultural production when they build a museum to assemble photograph albums, letters, and mementos gathered from an entire group whose shared experience calls for the display and sharing of collective memory (as with the case of the new African American History Museum in Washington, D.C.). Collections of cultural artifacts such as love letters are not only worth examining for individual family histories; they're critical to study as material *cultural assemblages*— objects that are grouped together because they have been classified as similar (Bennett 2007)—that tell the story about how culture works for an entire group of people.

Notes

1 In this book, I opted to omit a lengthy overview of how romantic love has been studied over the centuries, in part because it saves space for my new analyses, and in part because others have already written about this so extensively and effectively. For overviews of this centuries-old wrestling with defining romantic love, I recommend Ann Swidler's (2001) *Talk of Love: How Culture Matters*, Eva Illouz's (1997) *Consuming the Romantic Utopia: Love and the Cultural Contradictions of Capitalism*, Illouz's (2012) *Why Love Hurts: A Sociological Explanation*, and Carrie Jenkins's (2017) *What Love Is and What It Could Be*.

2 For an example of contemporary ethnographic epistolary work, see Katherine Carroll's 2015 article on the "dialogic epistolary" form of scholarly writing, where her research findings about the politics of donor breast milk are presented as a letter between milk donors and recipients in order to make the work accessible to those outside of academia, and in order to communicate in a genre that honors the emotional labor of field work and the subtle and affective moments of an intimate topic.

3 One example where this is present is in Teo's (2005) historical research about Australian love letters from 1860 to 1960, noting places in the letters themselves where writers referenced whether and where they kept their letters. Historians are increasingly likely to include curatorial practices in their investigations, if possible, as the move toward studying letters as cultural artifacts has gained popularity in that field of study.

2

THE DIGITIZATION OF LOVE

Technology and Communication Within Romantic Relationships

Iris Lee is a metadata specialist at the American Museum of Natural History in New York. In a 2017 *Bytegeist* podcast entitled "Talking Love Letters in the Digital Age with AMNH's Iris Lee," she shares the story of the spreadsheet she created that includes text exchanges between her and her partner from the early stages of their romance. She notes that this was before she had a smartphone, which today would include applications or capacities (such as screenshots) that could copy and paste text into another digital platform. And so, she spent hours and hours typing 283 text exchanges manually into a computer spreadsheet. As a trained archivist, she is invested in the saving of sentimental things, especially if they are located in precarious places, which is how she views an old cell phone that was damaged before she had a chance to enter the remaining text messages. As a metadata analyst, she values the ability to use a digital platform to aggregate and preserve what she defines as meaningful. She mentions in the interview that she wonders if there are others like her, or if what she is doing is unusual.

One of the inspirations for my research on love letters was a set of separate conversations I had with three friends a few years ago that may suggest that Ms. Lee is not alone in her desire to preserve messages of love, and not alone in her desire to do this using a digital platform. One friend, a woman in her mid-twenties, told me that she was saving texts from men she found attractive in a digital folder. Another friend in her early forties shared that she had created a folder in her phone's memo application to paste any electronic messages of endearment from her husband. A third friend, a woman in her late forties, shared the story about the box containing letters her college boyfriend (now her husband) had written while they were apart in study abroad experiences. In these cases, my friends wanted to preserve what they saw as sentimental artifacts from romantic relationships (albeit about relationships with different longevity and different motivating factors). And in all of the cases, they were educated professional White women saving messages from men with similar backgrounds. But they used different platforms and were different ages.

I am not the only one inspired to ask questions about the impact of digital technology on romance and romantic communication. In a 2017 blog post entitled "From Love Letters to Facebook Messages: Has Technology Ruined Romance?" Leah Perry ponders the impact of technology on romantic love by delving into social changes such as the substitution of delivering flowers with clicking a button to express interest, texting exchanges instead of face-to-face conversations in early relationship stages, and declaring relationship status on Facebook rather than announcing it in person. She also discusses how daily communication patterns have changed how identities and roles work in romantic relationships, since the ability to text any time and with ease may reduce "alone time and privacy separate from the relationship," at the same time it makes keeping things private even easier since people can meet online anytime quite discretely.

All of these stories could be read as intriguing simply because they highlight the importance of digital or physical platforms for communicating, and of storage locations for commemorating, which is the central focus of this chapter. But they contain something else that matters in any discussion of communication, technology, and relationships: how demographic characteristics shape access, attitudes, and behaviors surrounding love letters. Just as historical circumstances shape how people communicate and how they may view proper ways to do so, social locations—age, socioeconomic status, race, gender, occupation, sexual orientation, education level, and nationality, to name a few—play a large role, too. This chapter, then, introduces ways that digital or paper format matters in people's love letter curatorial practices. But it is also dedicated to an explicit introduction of how group inequalities and *privilege* (access to valuable resources that come about because of group membership) may shape these practices. I examine the digital "stuff" of romance as well as the differential access to, and use of, technology by group.

Romantic Relationships in a Digital World

Pen-and-paper letter writing has dwindled in recent years with the development of email and cell phones for texting (Haggis and Holmes 2011; Brandt 2006), and increasing numbers of people of all ages are finding themselves communicating with each other online (Smith 2015; boyd 2014; Fox and Rainie 2014; Turkle 2011). Commentators discuss the benefits and drawbacks of everything from e-readers (Weisberg 2011) to apps that offer couples virtual "memory boxes" for their privately exchanged digital love notes and mementos (Collins 2013), considering whether holding something in one's hand or reading it on a smartphone screen affects the meaning of the delivery and interpretation of the message. Much research is being conducted that gets at how people interpret information and communication technologies (ICT) in terms of impact on their personal relationships (see, for example, Ansari and Klinenberg 2015). Some findings point to a concern that ICT will replace face-to-face contact and lessen the human connection people have to each other (Turkle 2011), while others find the ever-changing technology landscape an exciting place ripe with opportunity for new forms of connectedness and innovation in relationship maintenance, especially with people who live far away, a possibility that used to be quite rare and challenging (Beck and Beck-Gernsheim 2014).

Any new research about romantic relationships must contain reference to how people use ICT. Online dating, with and without a desired path towards life partnership or marriage, is now less stigmatized and more common, with the Pew Research Center reporting a majority of people agreeing that online dating is "a good way to meet people," and with 5 percent of people in marriages or committed relationships saying they met their partner online. More than a quarter of adults ages 18–24 report using online dating sites or mobile apps, which has tripled since 2013 (Smith and Anderson 2016). Clearly there are growing numbers of platforms that cater to a variety of intimate partner desires, and growing numbers of people who use them.

Additionally, norms about how long to wait before replying to a text are discussed in popular films and television shows and in dating advice columns. There is no shortage of online references to that question, as exemplified by *BuzzFeed*'s quiz entitled "These Three Questions Will Tell You How Long to Wait Before Responding to the Person Who Just Texted You" (Hudspeth 2017), where one of the questions asks respondents to select the emoji that "fits the way you feel towards the person you're texting."

The role of ICT in romance doesn't end with meeting people through online platforms, though. The nearly ubiquitous access to online pornography and sexual imagery matters to couples who find these things to be troubling and to those who find them to be enticing and helpful. People continue to gain easier access to an increasing number of virtual reality and avatar-based online games that allow for sexual experimentation and romance, with or without a physically present intimate partner. There are more ways to communicate intimate pictures and thoughts, and sometimes save them to revisit later, in today's digital world. Every new version of smartphone includes fancier technology that can capture, organize, and share screenshots that can be used by partners to share intimate moments.

Indeed, there is an increase in the availability of applications explicitly devoted to preserving love notes, a trend especially connected to my research. For example, as Collins (2013) details, the South Korean app "Between" allows users to send each other messages, pictures, and digitized images, and it includes a faux-wood virtual "memory box" where couples can save their digital exchanges to look at later. The Google app "Digital Love Cards & Letters 2" allows users to create "unique" love postcards by customizing font, text color, and location of words and images to send to anyone. So, from swiping to clicking, from posting to sexting, and from creating an online dating profile to saving love notes in a virtual memory box app, the digital world is increasingly present in today's romantic world, and, in myriad new ways, romance has been digitized. As evidenced from these and other social changes, the form, frequency, style, and content of romantic communication have changed with the use of ICT.

Whether people believe digital communication to be problematic or helpful within romantic relationships, the fact of the matter is that it is increasingly part of intimacy. Gone are the days when sending a paper letter is the only way to communicate with a loved one who is not face-to-face. In order to look toward the future in terms of how current relationship mementos are preserved and understood, defining digital and paper love letters as cultural objects and asking about digital and paper love letter saving practices offers a helpful step.

Digital and Paper Love Letters as Cultural Objects

What seems more likely to get lost: an email saved in a cloud or a paper letter stored in a box in the basement? A person's answer depends on how they define a letter, and it depends on what their experience has been storing and losing items saved in different kinds of places. In both cases, there is a "thing" to be stored, though the object may be able to be held in one's hands (like a book with a cover and pages) or it may be able to be viewed only after clicking an "on" button and looking at a screen (like an e-reader).

A love letter is an object. I take that statement as a starting point, but I want to clarify that I use the term "object" to refer to either a physical or digital rendition of the letter. To understand whether and how digital and physical objects may come to mean different things to users, I apply Belk's (2013) discussion of the similarities and differences between digital and physical objects as extended selves and his framing of rituals of possession and disposition—what I call curatorial practices—of these objects.

In a 2013 elaboration of his 1988 theory of objects as extended selves, Belk suggests that "digital possessions are found to be almost, but not quite, the singular objects of attachment that their physical counterparts are" (490). When Belk first articulated his theory decades ago, he referenced solely physical objects. His updated conceptualization includes reimagining the self via digital objects. This work underscores questions about the impact of digital versus physical objects on identity and memory creation, and asks whether "a dematerialized book, photo, or song can be integral to our extended self in the same way as its material counterpart can be. If these items are stored on a remote server, are they really ours?" (479). Are digital versions of objects, because of their intangibility and easier reproducibility, therefore "less a part of the extended self" (Belk 2013: 481) than would be physical objects? Aligning with Appadurai (1986) and Denegri-Knott, Watkins, and Wood (2012), Belk articulates the possibility for individuals to "singularize . . . virtual possessions just as they can with real world possessions" (480). In this sense, there may not be much difference between virtual and physical objects as extended selves, and the meaning attached to saved digital and physical communication objects might not differ much from each other.

Whether the object is in physical or digital form can affect a person's practices surrounding that object. Despite his claim that people can attach meaning to both digital and physical objects, Belk (2013) also notes that virtual possessions lack a physical presence and can carry with them an uncertainty about control and ownership. Storing the text of an email love letter on an online server might seem safe in the event of a house fire. On the other hand, most individuals have a more difficult time throwing away a paper greeting card than they do deleting an e-card, at least emotionally (Siddiqui and Turley 2006). Further, because individuals tend to define digital family mementos as less valuable than physical mementos (Petrelli and Whittaker 2010), there might indeed be differences in how digital versus physical objects such as love letters are viewed, preserved, and visited. It may be that rituals of possession and disposition—the curatorial practices—of love letters vary depending on whether the letters are physical or digital.

In my survey research project, and inspired by Belk's choice to include digital objects as having potential to be considered meaningful, I define love letter in an inclusive way (see Chapter 1 for my definition and see the Methodological Appendix for how love letter is operationalized in the survey). I operate with the assumption that both digital and physical objects can be defined as meaningful, and that this can be assessed by examining digital and paper love letter curatorial practices. To interpret my findings, I incorporate ideas about whether there is something particular about love letters *as letters* that affects how their digital or paper format may come to be made meaningful as cultural objects. I rely on the masterful body of research and theory from cultural sociologist Liz Stanley, discussed most vividly in her 2015 article "The Death of the Letter? Epistolary Intent, Letterness and the Many Ends of Letter-Writing."

Stanley's (2015) exploration of whether the letter is now "dead" deals head-on with the impact of digital communication technologies on how people communicate with each other in writing when they're not physically in the same place. In addition to citing her own research on World War II love letters and emails from students, she debunks others' arguments about this "death." In doing all of this, she complicates the under-investigated and oversimplified claims that I've heard in my own research and writing that "the (love) letter is dead."

Stanley (2015) begins by outlining the argument posited by some scholars (e.g., Garfield 2013) that the letter is in fatal decline because of new digital technologies. Why would someone send a postcard on their travels when they can send a message in a social media platform? She pits this claim alongside the argument that new technologies such as tablets and smartphones (and their included applications for sending messages) afford growth of letter writing rather than decline, especially if the definition of *letter* includes fewer style and format conventions, and focuses more on the goal of the communication in the first place. In other words, the argument about the death of the letter stems from disagreement about what counts as a letter. If digital messages are included, love letters are growing. If not, they're dying.

The boundary of a letter has been challenged in many moments throughout history. In these historical shifts, three elements of letters are worth noting, as introduced by Stanley (2015): changes concerning the timing of letter exchange between people physically absent from each other; simulation of the presence of a writer; and the capacity for letters to be saved, which Stanley refers to as "the materiality of epistolary traces" (242). I discuss each of these in turn.

There's something about the rhythm of communication that matters. What role does not being face-to-face play in the definition of letter, especially if more or less time between exchanges matters in the definition of a letter as a letter?

> Absence is usually positioned as fundamentally definitional of the letter as a genre and is treated in the 'death of the letter' debate in a binary way: people are either present together in face-to-face ways or else separated from each other, with migrant letters viewed as the key exemplar.
>
> (Stanley 2015: 243)

But looking at people's private collections of letters, especially digital ones— whether a thread of texts in a smartphone or a folder of emails on a laptop— reveals that the absence is more akin to what Stanley calls "interrupted presence" (243). For most people who have loved ones with whom they communicate regularly and who live with them or nearby, daily face-to-face moments are bridged by small communications that occur when they're temporarily apart. So, the dichotomy of *synchronous* (at the same time) and *asynchronous* (not at the same time) communication is made messy. Time and space are compressed almost to the point that the exchange is instantaneous (or can be). Those who argue that digital communications are bringing about the death of the letter may claim that this bumps texts out of the definition of love letter. But, as Stanley points out, time and space are not dissolved when there are digital exchanges. They are merely made more rapid.

The idea of a love letter requiring time between exchanges, and thus the inclusion of digital notes as problematic in the definition, is predicated on the assumption that historical love letters have required time and distance. But, as Stanley elaborates, this is an ahistorical view. It has long been the case that newer and faster technologies used for communication (including, by the way, speedier transportation to get messages from one locale to another) have replaced older and slower ones. This was true when the telegraph and postcards came along. Importantly, "speed and time/space compression are not absolutes but are experienced relative to what existed before" (246). Further, as social researchers have pointed out for decades, when people begin new social practices, the newness of those practices is "accentuated, while over time they become assimilated into pre-existing modes" and may start to feel normal (250). To illustrate, every time I get a smartphone upgrade I acclimate to its new features, often with a bit of trepidation about whether I'll be able to handle the changes. But within a month, I'm used to it and have usually forgotten what the old phone was like. Stanley's research points to the notion that this can happen at a collective level, too.

Whether to include digital forms of communication in a definition of love letter is also discussed in terms of the significance of touch and presence, or how much of a person is imbued in the physical letter itself. A letter, whether handwritten or not, is a simulation of a person who is not present. But those who argue that the letter is dying suggest that handwriting, licking a stamp, touching the paper, and perhaps even offering their own perfumed scent on stationery offer a realness to a paper letter that cannot be replicated in digital form (Jolly 2011). It's as if there is less of a person present in a digital letter than in a paper one. Stanley notes, however, that many paper letters and notes do not contain these things, and digital communication increasingly adapts to contain elements of personalization such as emoticons, pictures, font color, or signatures.

Can someone keep, buy, sell, save, throw away, re-read, research, or give away a collection of love letters? The answer to this, according to those who argue that the letter is dying, calls to mind the importance of materiality: the form that a cultural object takes. According to this argument, a collection of love letters can only be preserved to have a "historical afterlife" (Stanley 2015: 244) if saved in physical form. But, of course, the digital world is filled with material things that can be touched and that leave physical traces, from phones to hardware to batteries to electrical systems in walls. And digital versions of things are quite durable. Even if an email is deleted, it does not mean it has been destroyed, as would be the case if a letter was burned in a fire. Thus, according to Stanley, the preservation of love letters can occur in either physical or digital format.

People write letters to each other on paper less than they used to, a finding that my (2015) and others' (e.g., Licoppe 2004) research has uncovered extensively. But this finding is based on a definition of letter that includes adhering to stylistic conventions such as "opening salutation, formal address of someone, message with standard syntax and punctuation and layout, pre-closing, and closing with signature" (Stanley 2015: 242). In other words, a short text would likely not count in this definition. Stanley suggests that digital communication may adhere less to these conventions than a pen-and-paper letter, but people should really look closely at the communication's *epistolary intent* (the aim of communicating with someone who is "not there" in either the same space or time as the writer, with the hopes that the recipient will respond) (242) and *letterness* ("the porous character of the letter and its ability to morph into other forms") (243) in defining something as a letter or not.

Using her own analysis of various types and formats of letters, Stanley argues that the goal of the letter (epistolary intent) is still the main element that constitutes it as a letter, and that new forms of messages (letterness) suggest growth rather than decline of this kind of communication. She cites numerous examples of letter-writing practices throughout history that challenge the expectation of physical absence, personalization, and materiality as previously discussed. Her examples negate the requirements of a firm boundary between a letter and other writing genres (as in de Beauvoir's grocery list example), an expectation of a response, and a supposition that a letter must be saved in the first place.

Most vivid in the sample of letters studied by Stanley are her parents' love letters from World War II, which she found in a jewelry box after her mother's death. These took the form of fifty envelopes with a single sheet of paper in each, sent to her mother by her father from various military locations and a prisoner-of-war camp. The content of each was short and consistent across letters: their names, drawings of hearts, mentions of the word *love*, and the symbols "XXX" to signify kisses for her and for their son. Stanley continues:

> The letter here was in a way tangible, but stripped to the bone, to elements my mother could read and write in return: a formal address she could copy, their names enjoined from one to the other, a message of continuing love expressed as hearts and kisses, and the repeated return gift of the letter. The past tense is used here, for they were cremated with my mother and both her and my father's ashes scattered together. I kept the heart, for love, and for remembering that while it may not be "a letter" it articulates the quintessence of epistolary intent and displays considerable letterness.
>
> (249)

In these letters, the writer intended them to focus on romantic feelings, they were part of an exchange, they were personalized, and they had materiality that allowed them to be saved in one place. Using Stanley's claims about epistolary intent and letterness, if she would have found a digital folder of text exchanges between her parents (if that technology were available then) that contained abbreviated notes, symbols, and reference to inside jokes or information that only an "insider" would know, I suspect she would say they would count as love letters, too.

Numerous moments when conventions of content and form of letters and other written documents have changed have taken place over time. There has also been innovation in both, whether it's creating new acronyms or symbols for affection, or in the norms about how formal a letter is supposed to be. Of course, love letters have changed, and this has been rendered especially visible given the vast availability of communication between people across time and geography. This change in letterness has included a requirement that time needs to be compressed. There is seduction in brevity. But there is also seduction in claiming that the love letter is dead, because to do so reinforces the cultural notion that romantic love must be communicated in written form, and must be represented with physical objects that are reinforced as meaningful in popular culture and economic exchange (Illouz 1997).

But if it is the case that there are many more forms of letters today (an increase in the variety of letterness), and if the intent of letter-writing (the epistolary intent) remains similar to the past, then it is hard to support the claim that letter writing is dying. Plus, while the time between communications has shortened, people have still not mastered the technology of being in two places at once. As long as there is physical absence, some form of simulated or proxy communication will serve as a bridge in a relationship, even if it's just to send a text during a workday when someone will see a lover in an hour. I return to the question of whether the love letter is dying in light of these points in Chapter 5.

Curatorial Practices Surrounding Digital and Paper Love Letters

In my survey I asked 373 respondents about their curatorial practices relating to their love letters (the survey design, questions, sample size and demographics, and analysis techniques are discussed in the Methodological Appendix). After asking whether they'd been in a romantic relationship where communication was salient, I asked them what types of communications from their relationship they had saved (with choices that included both digital and paper formats). I then asked them which type of communication they looked at most from these listed formats, as well as how often they looked at them. Following this, I asked why they usually look at that type of item and what motivates them to do so.

Most people who took the survey have saved some kind of communication from a past or present romantic relationship, but communications are not the only things they save. For people who have saved communications from a romantic relationship, nearly eight out of every ten people have also saved a physical or digital memento (e.g., stuffed animal or travel souvenir) defined as symbolizing the relationship that is not a piece of text-based communication. Respondents very much describe themselves as people who keep objects that are meaningful to them (62 percent say "This definitely describes me" and less than 1 percent say "This definitely does NOT describe me"). These findings reveal that, at least among those who took my survey, people are likely to save objects that are meaningful to them, an important characteristic to keep in mind as the rest of the findings are discussed.

I asked respondents to choose all of the types of communications saved from the romantic relationship they referenced throughout the survey. Strikingly, respondents overwhelmingly choose to save paper communications (68 percent save letters, 73 percent save cards, and 67 percent save handwritten notes) from their romantic relationships. Emails are saved by 47 percent of respondents, texting conversations are saved by 32 percent, Facebook message conversations are saved by 23 percent, captured Snapchats are saved by 6 percent, and other formats are saved by an additional 6 percent of the sample. Most of the romantic relationships referenced in this survey began in the last decade, so it is important to note that paper love letters are saved frequently even in the midst of a world where digital platforms are readily available.

Just because someone has a saved love letter doesn't mean they reread it. This could be because they do not know where it is or because they do not have a desire to revisit it. In general, people are not terribly likely to revisit their saved love letters. But if they do, they are most likely to revisit paper formats (23 percent revisit letters, 20 percent revisit cards, and 17 percent revisit handwritten notes). Texting conversations are revisited by 12 percent of the respondents, emails by 8 percent, Facebook message conversations by 5 percent, and captured Snapchats by 3 percent. Thus, people are more likely to save and revisit paper love letters than digital ones.

If someone does revisit a saved love letter, how often do they look at it? It turns out, not very often. Respondents primarily (67 percent) look at it a few times a year or less. Only 4 percent of respondents look at it daily or almost daily, and another 27 percent look at it between once a week and once a month. The revisiting of past love letters, for the most part, seems to be an occasional activity.

People are motivated to revisit love letters for a variety of reasons. When asked to check all of the reasons why a love letter may be revisited, respondents are equally likely to do so to feel nostalgic as they are to accidentally find it doing other things (50 percent in both cases). Nearly half (45 percent) say they revisit it when they are organizing objects or files where the item is located, and the same number do so to remind themselves of the good parts of the relationship. Just over a third (35 percent) revisit it when they are cleaning. Infrequently, respondents revisit a love letter to celebrate anniversaries or other special occasions (10 percent), or to remind themselves of what to avoid in a relationship (4 percent). So, people are as likely to revisit love letters by stumbling upon them while doing other things like cleaning or organizing as they are to intentionally revisit them to remind themselves (for better or for worse) about the relationship.

To summarize, my survey findings complicate the question of whether people still save love letters and what they may mean to those who save them. I find that love letter saving is alive and well, but that paper love letters are saved and looked at more than digital love letters, even as digital communication is increasing in romantic relationships. For both formats, the letters are revisited occasionally, with respondents noting similar likelihood to revisit them intentionally and accidentally.

By presenting these survey data, which will recur in the remaining chapters, I introduce a new way to study love letters. The curatorial practices surrounding love letters tell a more nuanced story than simply saying whether people save love letters at all. The findings presented here add to existing knowledge about how material culture matters in any understanding of communication within intimate relationships more generally. Since today's social interactions are in flux as a multitude of digital platforms enter the marketplace, it is important to investigate how one genre of communication may or may not be following those patterns. Ever-changing digital communication practices and norms may alter how people interact with each other. I offer that any investigation of how ICT is impacting relationships is incomplete unless people think about how communication formats matter in terms of curatorial practices.

Literacy, Letters, and the Digital Divide

Group Differences in Curatorial Practices

People have privilege if they have access to valuable resources, rights, or other advantages based on their membership in a group. For example, if someone is a student at a college or university and they have an ID card, they can access amenities such as a library, gym, or café that non-students cannot. This access constitutes a privilege that is based on membership in the group of enrolled students. Usually the term *privilege* is used by social researchers to refer to advantages that are not necessarily earned, but that come to people based on their group membership.

As the demographic information presented in the Methodological Appendix shows, my survey respondents are in groups that are privileged; they are disproportionately well-educated, mostly White, all in the United States, and earning incomes (or come from families that earn incomes) above the national median income. This is in part due to the distribution of the survey link (at least at first) within my social circles. Plus, the survey was electronic, thus favoring those individuals with internet access and enough technological savvy to navigate questions on their computers or smartphones. Survey respondents did vary by gender, they were geographically dispersed, and they represented all age groups from 18 to 86. So, I have responses from people who represent some variety in group memberships. But this has its limits. To alleviate the shortcomings associated with a sample that is not as economically, racially, or internationally diverse as it should be in order to be more representative, I offer here other people's research and larger contexts to frame my questions and findings in light of inequality and privilege.

A few of my findings relating to group differences in curatorial practices are worth sharing. There are not many differences in love letter curatorial practices between individuals in different racial-ethnic and social class groups. More specifically, in terms of race, 86 percent of respondents who were Hispanic or Latino/a, Black or African American, American Indian or Alaska Native, Asian American, Native Hawaiian, or Pacific Islander have saved love letters, while 89 percent of Whites have. In terms of education level, having at least a college degree means a slightly higher likelihood to save love letters, with 92 percent of those with a college degree or higher saving them and 86 percent of those without one saving them. Income does not seem to matter, with virtually identical likelihood (around 88 percent) to save love letters appearing across income categories. The biggest differences in curatorial practices emerge in terms of gender and age, which are the focus of Chapters 3 and 4. In Chapter 5 I return to the role of privilege more broadly in the conceptual model for understanding love letters, especially as it may relate to technology, communication, and the digital divide.

The Digital Divide in the United States

The idealized handwritten pen-and-paper love letter, as Stanley (2015) points out, is a largely wealthy, White, and Western construction, historically situated primarily in the 18th and 19th centuries. But a lot of time has passed since this construction. What role do demographics play in terms of access to digital communication technology today? One of my go-to locations for useful and accurate information about contemporary attitudes and behaviors about a wide range of social issues is the Pew Research Center. In their Internet and Technology section, they have a lengthy catalog of articles under the heading "*Digital Divide*," the term that refers to differential access to and use of ICT based on group demographics such as social class and race.

Today, with increasing access to digital communication platforms across demographic groups, one may presume that the sending and receiving of digital love letters (or any information or sentiment) has been democratized. Unfortunately, there is a dearth of research on how demographic groups vary in terms of their use of ICT in romantic communications specifically. So, a useful starting point is to understand general ICT access and use by group.

Access to ICT has increased across all socioeconomic status groups, but the story is complicated. On one hand, there are economic inequalities in terms of people's ability to access online news and use online tools to enhance their status, with lower income individuals having less access than affluent ones (Zillian and Hargittai 2009; Mitchell, Simmons, Matsa, and Silver 2018). Indeed, internet non-adoption is more likely among lower income individuals, those with lower education levels, older individuals, and those who live in rural areas (Anderson and Perrin 2016). On the other hand, there is near-ubiquitous possession and use of smartphones, laptops, and other ICT devices. Rainie and Perrin (2017) note that more than three-quarters of Americans say they own a smartphone, with adoption rates rising among lower-income individuals from 52 percent of households earning less than $30,000 a year owning a smartphone in 2013 to 64 percent in 2016; they also note that smartphone ownership is higher among affluent groups.

Perhaps more than access to the pieces of technology, it is necessary to examine how they are used. Regardless of technology access and frequency of use, class affects other elements of romantic communication, such as number of words used and frequency of communication. For example, Illouz (1997) found that working-class individuals in romantic relationships communicate less than their middle- and upper-middle-class counterparts, especially in face-to-face communication. This is because "verbal communication is part of the social identity of both men and women of the upper middle class" (279). Thus, while Illouz was not examining romantic communication in digital platforms, her findings suggest that ongoing research about how class groups may vary in the ways they communicate across platforms is warranted.

Most of the U.S. population is online, with only 13 percent of Americans reporting that they do not use the internet (Anderson and Perrin 2016). Rainie and Perrin (2017) identify few racial and gender differences in Americans' smartphone use, with 72 percent of Blacks, 77 percent of Whites, and 75 percent of Hispanics reporting ownership, and with 78 percent of men and 75 percent of women reporting ownership. There are some slight racial difference in inter-net non-adoption, with 13 percent of Whites and 16 percent of Blacks and His-panics reporting that they do not use the internet (Anderson and Perrin 2016). This means that class, gender, and race distinctions, while present, may be fad-ing in terms of technology use for everyday communication, although there is a lack of research on this in terms of romantic communication. Not only is it important to understand how people use technology beyond whether they have the equipment or how often they use it, it's also important to investigate how group status may impact the practices associated with communication—digital and otherwise—to understand the role of love letters in a changing world.

Changing Love, Changing Technologies, and Privilege in a Global Context

German scholars Ulrich Beck and Elisabeth Beck-Gernsheim (2014) articulate patterns associated with global influences in romance in their aptly titled chap-ter "Globalization of Love and Intimacy: The Rise of World Families." They begin by positing that what happens on the global stage impacts families and personal relationships on the smaller stage. Using the term "world families," the authors discuss how love and romance between partners (and also connec-tions between parents and children) operate in families whose lives are inti-mately connected but whose physical presence is separate by national borders or continents, a pattern that is increasingly common.

For example, family intimacy across borders is vividly experienced in *global care chains* (e.g., caretakers from one [usually poor] country, sometimes living apart from their own partners and children, caring for children from another [usually affluent] country). In these instances, the intimacy is maintained through geographic distance, and often includes the exchange of communi-cations back and forth. Increasingly, these exchanges are via digital platforms such as cell phones, suggesting that the digitization of romantic communica-tion across borders has allowed for a democratization of love letter exchange. However, whether a person has time, privacy, or ability to have and operate a cell phone depends a lot on the educational and economic resources that someone has. The role of digital communication can operate in different ways. On one hand, cell phones and international calling plans are getting more affordable, which makes communication between partners who are apart out of economic necessity easier. On the other hand, affluent couples (especially those from the Global North) who are separated can take for granted that they have the time, privacy, knowledge, and money to buy and use a device that keeps them connected.

Over the last two years, whenever I have mentioned to my anthropology friends and colleagues that I was writing a book about love letters, their nearly universal response was that I needed to look at Laura Ahearn's (2001) book

Invitations to Love: Literacy, Love Letters, and Social Change in Nepal. And so I did. In Ahearn's rich and thoughtful ethnographic exploration of how a Nepalese village changed from one with arranged marriages to one with many elopements, she underscores the importance of newly-acquired literacy present in love letter writing as a means to challenge old rituals and customs. She studied the content of 200 pen-and-paper letters, mostly written by men. It was through improved literacy and love letter writing that new courtship patterns emerged, suggesting that love letters are not merely symbols of how romantic love operates, but, as was evident with de Beauvoir's letters (Coffin 2010), they are tools use to effect social change. One important social change in the village was a shift from marriages arranged collectively to those that were valued in terms of individualization, romantic love, and freedom to initiate relationships with partners without the influence of extended family. So, in a shift toward the Westernization of marriage, love letters were used as a means to subvert tradition and move towards romantic love as a precursor to marriage. In the end, though, the paths toward a new kind of love within marriages that came about in part because of improved literacy did not necessarily undo the centuries' old traditions that prevented women from gaining social power. Modernization was viewed by villagers as a way to gain tighter emotional bonds and progress in terms of social roles. But with these changes also came negative impacts of globalization and development, which included women's loss of familial support if they opted to elope.

The love letters in Ahearn's research were written on paper, not on a laptop or smartphone in the form of a text or email. Ahearn studied the culturally contextualized content of love letters, which I do not. So, why include this work in a discussion of the ways in which format of communication and curatorial practices may matter in terms of privilege and social inequalities? The answer lies in the role that digital technology may play in rapid social transformation across the globe. If access to ideas, information, and tools (including literacy and writing devices) increases via globalized sharing of technology, then the disruption of social traditions, including traditional kinship practices, will continue. In the United States, people who are worried that digital notes may substitute for paper handwritten letters are participating in a process of social change, too. Do people hold on to the old ways or do they use new tools? And, are "old ways" and "new tools" necessarily mutually exclusive?

A recent article in *Brides* magazine (Emery 2017) explains that, while arranged marriage is still alive and well for around half of marriages worldwide, technology has altered some of the practices surrounding how the partnerships are formed. Websites catering to Indian families such as shaadi.com and indiamatrimony.com, for example, offer parents new ways to "advertise" their children as potential partners in their profiles, which, according to the article, is not a far cry from matrimonial ads placed in newspapers in the last century. Most salient for my research, though, are the changes that have come about in whether and how couples communicate before they marry. Now, digital technology allows couples to engage in conversation before they decide whether to marry more easily than in the past, often via digital platforms such as texts and emails—a pattern that is accompanied by greater likelihood for families to give their children an opportunity to veto the arrangement. The increase in cell phone availability and use has made long-distance courtships much easier (Voo 2008). This is most likely, though, among families with resources to engage in digital communication and whose daughters tend to be older and well-educated. For girls under fifteen years old who come from impoverished families, this choice and the communication platforms and opportunities are not available.

The 2017 film *The Big Sick* is about a young Pakistani-American man whose romantic interests (and career interests: he wants to be a comedian) do not conform to his family's wishes to arrange his marriage (and to have him become a lawyer). He is in love with a White woman who was born and raised in the United States; he emigrated from Pakistan with his family when he was a child. He fails to find an effective way to communicate to her (at first) that his family is attempting to arrange his marriage, and he fails to communicate to them (at first) that he is not going to participate in that traditional custom. In one particular scene, where he isn't sure about the outcome of his romance or of his girlfriend's health, he re-listens to voicemails from her. Although not written (and thus not part of the love letters that I study), his revisiting of these messages demonstrates a curatorial practice (he did not delete them; he saved them in a particular location; he accessed them not accidentally, but in order to remind himself of their relationship, perhaps nostalgically). The scene moved the story forward because it demonstrated the main character's move away from tradition and toward a modern version of love. At the crux of the dilemma for the main character was a tension between his desire to have the life his parents wanted for him—economic success, freedom, and a good life—and the traditions they were trying to uphold in the face of Westernization. Communication technology played a role in that the saved voicemails represented his pull toward a modern way and a push away from a traditional one. That they were saved, organized in one location, and accessible mattered in the character's desire to reminisce about his romance with someone he loved.

Beck and Beck-Gernsheim (2014), in their discussion about "love separated by geographical distances" (48), note that love remains abstract in the absence of a physical partner. The act of sending a love letter serves to "remind them of their moments of togetherness, and enable them to rediscover them, preserve them and reinforce them" (48). They also discuss at length the impact of long-distance love on a couple, both in terms of how they present themselves to each other, and in terms of what their relationship consists of:

> Distant love is the love of the Sunday self for the other person's Sunday self, purified of the mundane realities of ordinary life. It is a love in which you do not have to reach an understanding about household routines or the horrors of imminent family visits. But because you experience only aspects of your partner's life, and because you know about much in his or her life only through his or her narratives—in short, because many potential crisis zones are concealed by distance—the relationship lacks the connection to ordinary reality.
>
> (50)

If communication format is taken into consideration, especially in terms of time between exchanges, then the speed that digital communication affords may affect whether people are able to present themselves to each other as mundane selves rather than "Sunday selves." While I do not focus on long distance relationships, I do focus on whether defining something as a love letter may entail considering the pace (and therefore presentation of self) within the exchange. Perhaps paradoxically, the ease and speed with which digital love letters can be exchanged can afford couples not physically in each other's presence, even for an afternoon or a few days (as was the case in the scene from *The Big Sick*), to have a more real exchange. Further, because I allow for the love letter to be defined as something that can include other information (e.g., a grocery list), the mundane self can be easily present as compared to a letter that takes on the more traditional definition of pen-and-paper.

In my survey, I asked the following question: "Some people debate whether digital communication has replaced handwritten communication. In your opinion, do you think letter, note, and card writing, in handwritten form, is fading in romantic relationships?" Around nine out of every ten survey respondents said yes, a response that was consistent across racial-ethnic, age, gender, education, and income groups. When asked to elaborate about the assets and drawbacks of both digital and paper love letter writing in an open-ended question, respondents were likely to list assets of digital love letters such as "ease, speed, ability to converse in real time," (Janning and Christopherson 2015: 255), and the ability to be as meaningful as handwritten letters. Drawbacks of digital love letters included being impersonal or lazy in their tendency to promote instant gratification. When it came to paper love letters, assets noted included "tangibility, sacredness, rareness, and tendency to promote thoughtful and personal communication" and drawbacks included that handwritten communication was "cheesy, hokey, old-fashioned," and not necessarily more meaningful than digital communication (Janning and Christopherson 2015: 255).

Today represents an interesting communication crossroads with a few possible outcomes over the coming years. Perhaps there will be an increasing number of disappointed people who wish their partners had not replaced their stationery with an iPad because iPads are impersonal platforms. Perhaps there will be a backlash against the influx of ICT into romantic lives whereby people return to an idealized nostalgic past of handwritten letter writing and saving, as some younger people are doing, and which is discussed more in Chapter 4. Or, perhaps, as in the scene from *The Big Sick*, there will be a redefinition of love letter as technologies bridge people's desire for love and their preference for easy access to recall that love at opportune moments.

From the impacts of literacy among villagers in the developing world writing love letters to each other to the contemporary global movement of couples finding work and trying to communicate with each other, global issues of inequality and privilege matter when it comes to romantic communication between partners. This extends far beyond group differences within the boundaries of any single country.

Conclusion

In late fall 2017 I was interviewed in a podcast dedicated to uncovering whether, why, and how people may save their text messages from their dating relationships. In this interview, I responded to one of the hosts who talked about how it feels different to save texts as compared to saving a stack of love notes piled up on a desk. She saw herself as not wanting to save the texts, and she wondered if the process for deciding whether to save the stack of paper love notes would have been the same.

My response was that the process of deliberating about whether to save them may be the same in both instances (should I keep them or throw them away?), but also that the format may make the workload associated with saving, as well as the location for saving, differ. Curatorial practices for managing thousands of text messages look and feel different (emotionally and bodily) than managing a box of paper letters. Maybe it makes other things differ, too, which is why this chapter investigates whether love notes on paper may be somehow different from those in digital form. Or maybe perception of whether saving paper letters or digital ones differs depends on how someone defines love letter. This definition could be impacted by someone's views as to whether digitizing it changes its level of personalization, the pace of exchange, and the ability to preserve a love story using a platform viewed as more or less precarious. These

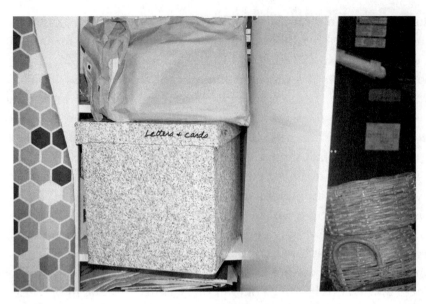

Figure 2.1 **Letters and Cards Box** (Source: Author Photograph)

perceptions may be influenced by group statuses. Maybe, for example, some-one's fear of losing a collection of love letters depends not only on their per-ceptions of the security of digital clouds or their basement closet, but also on whether they grew up in scarcity. Likely their view is shaped by how closely they can adhere to culturally prescribed ideals of romantic love that are impacted by economic status and global location, or how much they have access to digital forms of communication.

Around the time when my three friends shared their stories with me that inspired my research, as discussed at the beginning of this chapter, I stumbled upon a box in my basement labeled "Letters and Cards" (see Figure 2.1). I had covered the box in contact paper to make sure it looked more important than other boxes that may accidentally get thrown away. One weekend afternoon I brought the box upstairs and sifted through the oodles of paper letters from friends, family members, and past romantic partners, all from my high school, college, and graduate school years. My husband walked into the living room, finding me in a sea of spiral notebook scraps and folded papers, and wondered what I was doing. I told him and shared a few snippets from some particularly hilarious letters. Pausing, and realizing that both of us had entered college before there was email beyond intranet (a digital network within a community) and left college around the time that the World Wide Web was in its naissance, he proclaimed, "I think we may be the last generation of letter writers." He said this with a particular definition of letter in mind—one that, as Stanley (2015) notes, some say is not possible in digital platforms because they compress time to the point that they don't count as love letters. While I don't think he meant it that way, his comment did make me wonder whether the classification of something as a love letter is in the midst of being reconsidered. In the end, I discovered that including digital platforms in my definition of love letter was the right call on the survey, particularly because the epistolary intent was the same: to express via written communication a romantic connection to someone. Whether on paper or in a text, a message of love can be short, long, formal, informal, personalized, depersonalized, and embedded in other text that may or may not include a grocery list.

To send and read and then find and re-read a paper love letter has one kind of rhythm, while copying and pasting and digitizing and then re-reading a digital love letter has another. Like two dances with different rhythms, these letters take on different formats—formats which may affect their content, style, and overall message. But they are still love letters like the two dances with different rhythms are both still dances. As my response to the podcast interviewer suggests, while formats may differ in a love letter's letterness, the motivation to communicate romantic love (epistolary intent) may transcend format. Same motivation, different format. Same desire to dance, different dance moves. Or, as blogger Leah Perry (2017) decided, "maybe romance has not been ruined, but redefined." My survey results reveal that paper is the preferred format for saved love letters. As the next chapters discuss, it is important to explore what happens when what people want a love letter to be may not necessarily be what a love letter is likely to be, given greater participation in digital communication to send a message of love.

3

SPACE MATTERS

Where and How Love Letters Are "Curated"

In the previous chapter I noted that my husband's comment about people being in their mid-forties perhaps being the last generation of letter writers put the final touches on my motivation to do this project. As my excitement about the research grew, it dawned on me that the letters and notes from him, which began in graduate school, were not in the carefully contact-papered "Letters and Cards" box I had found in our basement. But where were they? After searching the house and wracking my brain, weeks later I realized they were in my dresser drawer. Apparently, the love letters from the one I loved most were placed in a different spot than those from other people, suggesting that location of saved objects is a salient part of someone's curatorial practices. Where someone places cherished (and not-so-cherished) objects, and why, is the subject of this chapter.

Where cherished possessions are stored matters, but so does whether people *know* where they're stored. The desire to know where something is located is impacted by how cherished the objects are. People can be uncomfortable if they don't know where things are, which happens to me every time I can't find my keys. Without keys I feel as if I have lost a tool to function instrumentally in my life. I cannot drive or unlock my doors. I'm not terribly saddened or worried, though, if I cannot locate a newsletter that I know I'll probably recycle anyway. But what about not knowing where a love letter is? This is where the expressive function of a saved love letter matters—its emotional importance. When I could not immediately recall where my husband's letters were, I felt unsettled.

I was recently told a story by a young woman who had no idea where the love letters were from her high school dating partner. This was particularly sad, she shared, because this person had recently passed away. In contrast to this, another woman told me about the relief she felt after the boxes of her past romantic partners' love letters had been destroyed in a plumbing disaster in her house. But her spouse's letters, she said, were missing, which brought her a feeling of unsettledness instead of relief.

These stories certainly call attention to the location of stored love letters: a basement, a box, a dresser drawer. But they also call attention to people's reaction to whether they know where they are, and the reaction to whether they have control over their safe storage. Both of these relate to how people view the relationship that yielded the love letters in the first place, a view that is shaped in part by cultural values. Put more concretely, the young woman who had lost her now-deceased high school partner's love letters could not hold on to something that contemporary culture says should be preserved—the story of those who are gone. The woman who felt uneasy about not being able to find her spouse's letters was not adhering to the cultural expectation that she should know where the items are that are from people who matter.

Making sure people know where something is means it is considered important enough to know about. For the second woman, knowing that past dating partners' letters were destroyed both lessens the need for her (the possessor) to feel responsible for their demise, as well as adheres to norms about getting rid of things from people who no longer matter. Not knowing where something is that one is supposed to know may make someone feel as if they're violating what's normal in society, which likely is why she felt uneasy not knowing where her spouse's letters were. This is particularly salient for love letters, since they call to mind boundaries between public and private, and between people who matter and people who don't. They capture the importance of objects that call to mind emotional experiences and they capture larger cultural values about what emotions are supposed to be symbolized by cherished objects. Where people put things can show what those things mean.

In these stories about the two women, I used the terms *partner* and *spouse*. I did this on purpose to allow for the imagined partners to be unclear in terms of gender and sexual orientation. How aware and interested someone is in the concept of *heteronormativity*, defined as "a system that works to normalize behaviors and societal expectations that are tied to the presumption of heterosexuality and an adherence to a strict gender binary" (Nelson 2015), may impact their likelihood to assume the partners were men or not. In other words, an assumption that the referenced relationships were heterosexual would demonstrate adherence to heteronormativity. This is because a heteronormative approach assumes a gender binary (where partners are one of only two gender categories) and assumes a monogamous heterosexual partnership between two cisgender—those whose sex matches their gender identity—individuals.[1] I aim to show how gender and sexuality matter (Rutter and Schwartz 2011) in discussions of love letters through the examination of the importance of spaces and places where people keep their stuff.

How people arrange their stuff, including where they put it (and whether they have room), can tell a lot about individual preferences for memory-keeping and the definition of self. But it can also tell about values that are dispersed culturally—values about privacy, storage, whose stuff is supposed to matter, how much stuff is too much or not enough, and how all of this is wrapped up in expectations about gender and sexuality in relationships. Since these expectations can be limiting and create inequalities, I examine my survey findings with an eye towards critiquing the heteronormative expectations surrounding love letter curatorial practices.

In this chapter, I discuss current attitudes and behaviors surrounding the storage and management of stuff in domestic spaces. I share my survey findings about where people store their love letters and what that may say about those letters' meaning to their possessors. I highlight gender differences, too, noting the significance of romantic love as a gendered experience and connecting gendered and heteronormative love letter curatorial practices back to how the arrangement of stuff in homes matters.

Storage, Tidying Up, and Too Much Stuff

In the January 2018 issue of *Real Simple* magazine, an article entitled "Project Declutter" by Petra Guglielmetti included tips for how to deal with "sentimental clutter" such as a grandmother's china set or a child's treasured toy. The advice about saved letters referenced in the article was offered by Beth Penn, author of *The Little Book of Tidying*, who advised determining "what's worth the real estate. Keep only those greeting cards with a meaningful note, not just a signature" (95). This advice comes on the heels of an explosion in the marketplace of books and websites devoted to "tidying."

Most notable in this marketplace is Japanese cleaning consultant Marie Kondo's 2014 book *The Life-Changing Magic of Tidying Up*, which has sold millions of copies worldwide and pairs with her acclaimed cleaning/organizing consulting business. In this book, Kondo advocates decluttering by category of object in the home, whether shoes, dishes, or greeting cards. The assumption here is that *type of object* is a real, useful, and boundaried form that is used to structure a project meant to alleviate stress and induce a calm sense of well-being by removing the visual reminders of the stress. Kondo advises people to ask themselves as they pick up every object within a category, "Does this spark joy?" If it doesn't, it gets tossed. If someone was to follow her guidelines, they would not pick up a love letter and a pair of shoes and decide which one sparks joy more than the other. This is because there are different values placed on categories of objects. Shoes can be about comfort and utility, and they can also be about style and aesthetic pleasure. Certainly decluttering surrounding shoes often has to do with getting rid of pairs that are more about emotional affinity than practical utility. But love letters are more likely to be defined categorically as expressive, not instrumental. Wants, not needs.

What's interesting about Kondo's ideas is that joy, like love, is presented as an emotion. As delineated in Chapter 1, emotions are part of an individual's sense of self (e.g., these shoes make me feel joyful), but they are also embedded in a cultural system (e.g., this year, loafers are in style so they are more likely to make me feel joyful because I am adhering to what's accepted or popular). In this case, tidying up and getting rid of things that don't "spark joy" must be situated in a social context where minimalism, decluttering, (at times) fear of environmental degradation, and calm in the face of hectic lives are so valuable that they can induce radical behavioral changes among those who take her advice.

The *Real Simple* advice column about getting rid of sentimental objects refers to storage places in the home as "real estate," as if they have monetary value in a housing market. The target market for this magazine, and for Kondo's book, is women. Despite the framing of storage spots as valuable "real estate" within a home, domestic spaces as a category are devalued relative to public spaces or workplaces. This devaluing is, in part, because they are still feminized realms. Even with strides in the paid work world and politics, and with more men taking on home-related responsibilities such as child care, cooking, and cleaning, women are still socialized to be the primary bearers of responsibility for the home, including its tidiness, organization of possessions that signify family roles, and décor (Janning 2017). This responsibility has been framed (and experienced) as burdensome, a result which has yielded a cadre of self-help books meant to aid (mostly) women in the management of a place that is increasingly hard to manage. Since the storage of love letters calls to mind both the preservation of family and relationship stories, as well as the labor required to organize and store them, it is important to see not only what people's love letter curatorial practices are in terms of storage location, but also to see whether gender matters in these practices.

Where Are the Love Letters, and Why Does Location Matter?

Objects can be denoted as special through curatorial practices. If someone keeps something in their possession and looks at it often, it can be defined as meaningful. But it's not that simple, since looking at something infrequently doesn't necessarily mean it is unimportant. Relatedly, getting rid of something and never seeing it again can signify the object as just as meaningful as keeping something and revisiting it often. That's why asking people about what they're doing when they revisit an object, which gets at why they revisit it, is important (recall from Chapter 2 that people are as likely to revisit a love letter intentionally as they are to do so accidentally).

But whether someone revisits an object depends not just on their desire or motivation to do so, it also depends on whether they can find it. Indeed, where people put things can also show how (and how much) they matter—how an object, as Epp and Price (2010) discuss, becomes meaningful through a process named *singularization*. If something is singularizable, it is decommodified (made meaningful apart from its monetary value), made more emotionally attached, and given personal meaning (Epp and Price 2010; Miller 1997). Although directly asking someone if an object is meaningful can illustrate something about its singularization, the location of the object might further reveal something about its meaning to its possessor.

The location where an object is stored can be designated as either *heated* or *cooled* (Epp and Price 2010). Putting an object in a space that is easy to see or access is heating the object. Putting an object in a space that is hard to see or access is cooling the object. Heating and cooling can tell the story of the object's significance to its possessor. For example, keeping an object in the middle of the living room where a family spends a lot of time will heat the object, making it more visible, salient, and often more readily and frequently seen or touched by family members. For an individual, an object might be heated not necessarily by being accessible in more public domestic spaces such as a living room, but by being placed in a spot that is easy to see or reach without needing to open a door or drawer (such as on a nightstand, desk, or wall).

Conversely, storing something in a cellar that is far from other living spaces or in a closet or drawer that hides its contents from the sight of household residents will cool the object (Epp and Price 2010). This happens when someone wants to get rid of, say, a coffee table. A first step is to put the table somewhere such as a basement or garage to see what it's like not to access it. After cooling the table in a less accessible spot, it may be easier to get rid of. Accordingly, one could argue that the cooling of a less meaningful object occurs because it is not easy to access without opening something or removing a barrier to make it visible, and therefore that object might be less salient to an individual's self-definition. The cooling of objects often signifies a distancing of those objects from someone's definition of self. "Transition places" such as basements, porches, or unused rooms in the periphery of a home (as opposed to core spaces such as kitchens, living rooms, and bedrooms) are often used as locations to isolate possessions in order to move them "away from the domain of 'me'" (817). In this way, cooled objects may be defined as less meaningful because of their less-accessible physical location within a home.

What may matter for the storage of love letters, as opposed to something like a table, is that they are defined as more private and personalized objects. Love letters may embody a person more than a store-bought table does; unless a table is hand-made by a love one, love letters are more likely to contain evidence of bodily action such as handwriting from a known person. So, perhaps making them less accessible (that is, in cooled locations) could construct their meaning as sacred and private, and therefore more meaningful than if they were in heated locations. This could complicate the notion of heated and cooled locations as more and less meaningful depending on the nature of the object being singularized, at least in terms of physical objects.

In an open-ended question on my survey, in order to capture how the love letters may be singularized in terms of location, I asked respondents where they store their love letters. The most prominent theme in the open-ended responses to the question of where paper love letters are stored references whether they are in hidden or displayed locations, which align with Epp and Price's (2010) cooled and heated locations, respectively. If a respondent says that their love letters are in, under, or behind something, they fall into the hidden category, representing cooled locations. Love letters that are positioned horizontally or vertically on something are in the displayed category, a heated location.

I also classify the responses by whether people refer to paper or digital storage formats in terms of their storage location. Unfortunately, in the coding of digital communication patterns, the open-ended responses do not contain enough details to code whether they are heated or cooled locations as defined by Epp and Price (2010), since the physical location of the digital device is not noted, nor are details about how hidden or visible the digital communications are.[2]

Since the most frequent formats of saved love letters for these respondents are paper (letters, cards, handwritten notes), it is not surprising that the most common emerging theme from the open-ended responses about storage does not contain reference to digital locations. About half of the respondents say that their love letters are stored in physically hidden places (e.g., "in a drawer," "under my bed," "behind a picture," or "in a decorated storage box"), and just over a tenth of the respondents note that theirs are stored in physically displayed places (e.g., "on my bulletin board," "on my desk"). To borrow Epp and Price's classification, respondents choose cooled (hidden, hard to access) storage locations more than heated ones for their saved paper love letters.

In terms of digital storage, which is less common because paper love letters are more likely to be saved and stored, about a fifth of the respondents store digital love letters in email folders or on their computers. Another tenth store them in their smartphones. It is difficult to tell whether computer or phone storage represents a heated or cooled location, though, since the survey responses do not include descriptions of how the letters look or where the phone or computer is physically stored.

To recap, love letters (especially paper ones) are likely to be stored in places that are hidden and not immediately accessible. Being stored in a cooled location, for Epp and Price (2010), means it may become less singularized to the owner, but my analysis suggests that keeping something hidden (in a cooled location) not only means it may be looked at less often, but also that it may be rendered even *more* meaningful because its hidden location can connote a private sacredness. And it may be that the idea of hidden can vary throughout the home. This possibility hits close to home for me, in part because I store my husband's love letters in a dresser in my bedroom (accessible but hidden in a drawer) and others' letters in a box in my basement (less accessible and also hidden behind a door). It also resonates with me because, when my father died, I snuck a small four inch tall pink statue of a mouse into his coffin. The mouse, cartoon in its features, had its arms spread wide with the caption "I love you this much" typed across the base. My dad's nickname for me was *Die Kleine Maus* (the little mouse), a German term of endearment often reserved for children. I put the statue in the coffin not because I wanted the reminder of my dad's love for me (and my love for him) to be hidden and thus less meaningful. Quite the opposite. I did it precisely because removing it from circulation rendered it even more meaningful: so precious that it cannot be touched.

Perhaps in a parallel way with regard to the dispossession of love letters, for some people, the act of burning love letters, a practice that I was not able to capture with my survey research on people's saving practices but that has been shared with me as people tell me stories of their own curatorial practices, demonstrates an extreme measure to keep them private and thus more meaningful. Destroying something so that nobody else can access it can define the object as special, either because of a ritual that may cathartically process a separation or loss, or because of a desire to keep it so private that nobody can ever see it. Sometimes this has to do with believing that hiding something makes it more special or sacred; sometimes it has to do with worrying about whether the contents, which are defined as private by their possessor, would be shared with a wider audience, thus leading to embarrassment.

Because most people save paper love letters in my research, my findings do not shed much light on how digital spaces matter in people's curatorial practices, especially in terms of access, privacy, and the distinction between heated and cooled locations. For example, consider the possibility of someone having a folder on their computer desktop labeled "Tax Documents" when it's really a bunch of steamy love letters sent via email and saved as screenshots. Would this be in a heated location because it's close at hand, accessible merely with the double-click of a mouse? Or would it be in a cooled location because the folder name serves to "hide" the contents?

Despite my inability to capture how digital places operate in love letter cura-torial practices, my findings do reveal that, perhaps, objective measures of which storage location is easier or harder to access—a computer folder or a box in the basement—are less interesting than people's *perceptions* of how accessible and/or private a space is. It may be, for example, that one person perceives a string of text messages stored in a smartphone to be easily accessible because their phone is with them at all times. Another person, in contrast, may perceive this set of stored communications as inaccessible, because to find one particular message requires scrolling through hundreds or thousands of messages. Sometimes, then, the belief that something is important may depend on how someone may perceive the ease of access that is related to how the collection of messages is organized.

Further, I am interested in how these perceptions may be based, at least in part, on group membership. For example, the belief that a computer folder is easier to access and less private may be more likely held by those who are familiar with the security of digital file storage, or by those who have experi-enced a disaster that ruined their physical possessions. Just as growing up in a time period or family where scarcity was ever-present may make someone have a harder time getting rid of things they perceive to be important or useful, how precarious a place is perceived to be in terms of its security from the wandering eyes of others depends on factors such as whether someone was socialized to believe that digital storage is trustworthy.

Of course, how people individually view privacy in terms of love letter storage may matter as much as, if not more than, the format itself, and it may matter more than their social surroundings and collective messages about whether pri-vacy is valuable or not. If someone is a private person, they'll find a way to hide the letters no matter what, whether it be in a cleverly named desktop folder or an unmarked box in a closet in the basement.

Gender, Heteronormativity, and Love Letter Storage

In Chapter 2 I noted how demographic groups differ in their access to liter-acy and digital technology and in their use of romantic communication, with an emphasis on economic inequality, race, and experiences that showcase how globalization, immigration, and Westernization impact romantic relationships. Here I expand my survey findings about the location of stored love letters by focusing on gender and sexuality. I do this because my findings about men's and women's curatorial practices concerning the location of stored love letters tell the larger story of how gender operates both in heteronormative conceptions of romantic love and in the everyday practices that make up household division of labor. In other words, where men and women store their love letters in their homes gets at gendered cultural values surrounding romance and family roles.

To start, it is important to note that gender does not seem to matter in the responses to the question about whether handwritten letters, notes, and cards were fading away, with around nine out of every ten men and women answering *yes*.[3] But what did their own behaviors say about gender?[4] In terms of the curatorial practices introduced in Chapter 2, I found that men and women are similar to each other in their greater likelihood to save paper (around 66 percent) over digital love letters (between a third save texting conversations or Facebook message conversations, and about half save emails). They are also similar to each other in terms of why they may revisit a saved digital or paper love letter: both are as likely to "stumble upon" the letter as they are to intentionally revisit it.

Men and women differ along a few lines, however. First, women save more mementos and romantic communications from relationships than men (scoring 4.5 as compared to men's 4.1 on a 5-point scale asking them to rate how much the statement "I keep objects that are meaningful to me" reflects them, where a five indicates "this definitely describes me"). Second, men revisit the love letters that they've saved more often than women, despite the finding that women are more likely than men to save love letters. And third, men and women differ in the location of their stored love letters. These gender differences in curatorial practices relate to the place where letters are stored, and are elaborated in the subsequent paragraphs.

Men, while less likely than women to save love letters, look at their saved letters more frequently than women. Why might this be, how might this finding relate to storage location, and what might it suggest about how gender operates in romantic relationships and society more generally? Where someone may store a love letter affects how accessible it is, which may impact how often it is reread. The location of stored love letters differs by gender. As I already revealed, most people store their paper love letters in hidden locations (in, under, and behind something else). However, men are slightly less likely than women to do so (just less than half of men and over half of women), and men are more likely than women to display their love letters in more accessible or visible spots (on something else) (16 percent of men and 6 percent of women).

In addition, when I looked at the details in the responses pertaining to what type of storage container is used, only women mentioned a detailed articulation of what the storage container looks like (for example, that it is decorated, made of a specific material such as fabric [or, as with my box, covered in contact paper to make it seem less disposable], or is described as "special"). These findings reveal that, using the definition of a cooled object as one that is less singularized by virtue of it being placed in a less accessible place, it appears as if women singularize their paper love letters less than men. Women's love letters are more likely to be hidden and less frequently revisited than men's. However, the fact that women are more likely than men to offer details of the specialness of the storage container may suggest otherwise. It may, in fact, suggest that heated and cooled locations are not the only ways to discern whether an object is singularized to its possessor, and that locations are not uniformly defined as hidden or not.

These findings tell contrasting tales that are found elsewhere in social research about gender and romantic love. On one hand, the accessibility and visible display of love letters for men, and their frequent revisiting of them, aligns with claims that men may be more romantic than women, thus dispelling traditional claims that women are more invested in the emotional aspects of intimate relationships. For example, Illouz (1997) finds that "men are more likely than women to hold a romantic view of relationships" (219) among her interviewees, claiming connectedness based on affective dimensions such as "love at first sight" or simply "falling in love" without any explanation, rather than on a list of traits that add up to a desirable romantic partner.

On the other hand, my survey findings may not disrupt traditional gender roles as much as they support them. But even this interpretation is complicated. First, women are charged with maintaining kinship ties, which can take the form of interactions with people to stay connected or which can take the form of managing the material objects that symbolize those ties. Consistent with past research uncovering the gendered tasks of creating and organizing family photographs (Janning and Scalise 2015), women feel more pressure to preserve familial relationships as part of their role within families. Part of this preservation is securely keeping and storing cherished objects such as photographs and letters that symbolize familial relationships.

In addition to pressure felt by women to maintain relationship ties is the pressure to keep home objects organized and put away "in," "under," or "behind" something. My research findings may demonstrate women's likelihood to tidy up and men's likelihood to leave their belongings "on" surfaces instead of putting them away, rather than the location signifying how important the objects are to their possessors. Home magazines, self-help books, and the decluttering industry are marketed primarily to women. The storage of love letters in places that are "put away" fits well with the messages that women receive about being the managers of household clutter. What does not quite fit with this interpretation is the notion that women, in addition to feeling pressure to preserve personal and familial relationships via material objects by storing them in put away places, also feel pressure to share family stories with other family members to ensure that the stories are preserved over time. This might lead to the conclusion that women would be more likely to store their love letters in heated locations, making them more accessible to others. But women don't seem to do this. That women's love letters are hidden objects does not lessen their meaningfulness. In fact, it may be the opposite. The storage of love letters by women in cooled locations may actually signify their attempts to be a good (heterosexual) partner and, for those who have children, to be a good mother. To declutter and put things away is to curate the stuff that signifies a woman's successful role as wife (to a husband) and mother.

What may complicate this interpretation even further lies squarely in the understanding of cultural expectations for women in terms of heteronormative displays of sexuality, which aligns my survey findings with heteronormative expectations about gender roles. If women are supposed to preserve familial ties and adhere to their prescribed role of wife and mother by securely storing (and saving more) objects that symbolize those ties, men are more likely to be socialized to "show off" their romances. In this sense, the heated locations of love letters for men may serve as "trophies" of their romantic (in terms of sexuality, not in terms of marriage or parenthood) successes. My interpretations are tentative, however, because my survey findings cannot capture the motivation for why particular storage locations were selected.

Nonetheless, it is important to consider whether the gendered dynamics of romantic relationships may play out in love letter curatorial practices. What is expected from men and women stems from a system of heteronormativity that reinforces women's likelihood to feel responsible for family relationships and men's likelihood to desire sexual success (especially with women) (Nelson 2015). What my findings reveal is that this may be present in love letter curatorial practices, but further research is warranted in order to find out whether the motivation for particular storage locations differs by gender, and whether there may be different findings among those respondents who identify as homosexual or who do not fit into a gender binary in terms of their identity. In addition to this, future research should pay attention to more specific digital locations (i.e., not just with an open-ended question about storage location that yielded results such as "in a computer folder"), especially because there may be gendered patterns that relate to the privacy, security, or perceived tidiness of these places that differ from those I found in people's curatorial practices surrounding paper love letters.

The questions of where people store love letters and how private or accessible the storage place is, need to be examined by considering gender roles. Women are socialized to be the preservers of kinship relationships, whether via photo albums, inherited china sets, or love letters that capture a moment in time that matters for a particular family's story. That might explain women's greater likelihood to participate in research on topics that center on interpersonal relationships, as evidenced by the fact that most of the people who took my survey are women. But because most of the place-based storage practices I studied relate to paper love letters, I am not able to discern whether gender may matter in terms of digital versus paper love letter curatorial practices. I am also not able to discern whether content affects storage location. These limitations matter. For example, it's one thing for a woman to save a love letter that she is willing (and perhaps aiming) to have passed on to children and future grandchildren as a memento of a romance that played a big role in their family's story. It is quite another to have her sexually explicit texts found by a teenage child as they use her phone to order a pizza. Here, then, it is important for future researchers to uncover the overlapping relationship of epistolary intent, format, and curatorial intent to the performance of gender and sexuality in families and intimate relationships.

Conclusion

When I was interviewed about my love letters research a couple years ago, an article about the research appeared in a Chicago newspaper. After its publication, I received an email from a woman around my mother's age who had a collection of her family's paper love letters. She had no children or grandchildren, and no local or regional archive or museum wanted them. But, she wondered, was it okay to just throw them away? To her, she told me, that felt irresponsible, disrespectful, and sad. I wrote back to her and said that it was, of course, up to her if she wanted to get rid of them. I acknowledged that she wanted the letters to be stored somewhere important, to preserve her family's love story for an imagined future audience. I also acknowledged that if she needed permission to throw them away, I was comfortable granting that permission. It wasn't up to me, though, and there was nothing in my research that would suggest there was a right way to handle all of this. She didn't respond to my email.

Three weeks later, a package of her love letters arrived at my office, tucked tidily into plastic sleeves and accompanied by a note of gratitude. They are still sitting in my office file cabinet, unopened by me. I have a small sense of feeling responsible for protecting these, for fear that if I threw them away I'd somehow be disrespecting her family's love story, especially since she didn't have children. They have been singularized by her because they are in a place that she has labeled as safe and important. I believe that she wants me to singularize them, too. But in a way, it doesn't matter what I actually do with them. I could throw them away, open them and share scanned versions of them on Facebook, or reorganize them into a book and send them back to her. What matters is that she perceives them to be in an important place.

So, it seems as if responsibility for preserving something that ought to be respected and honored is at the forefront of people's minds when they

deliberate about keeping or throwing away love letters, and when they deliberate about where the love letters ought to be kept. Where the letter is may vary
from person to person. That the letter may be near or far away can tell part
of the story of how singularized they are. But, as this woman's story shows, it's
more complex than this. Perhaps the actual location of where love letters are
stored matters less than someone's perception of the location as appropriate.
This, as I return to in Chapter 5, depends on whether a love letter is defined as a
physical object that needs to be accessed in order to define it as a love letter per
se. It also depends on whether someone thinks digital and physical locations
differ in their ability to singularize an object.

What will be interesting to see in the future is whether digital love letters
become more likely to be saved than my 2013 survey responses show. These
days, five years is an eternity when it comes to technological change. Future
research may uncover more nuance about curatorial practices, as well as singularization, relating to digital love letter storage. Perhaps increasing use of digital love letter storage, as with the virtual "memory box" app referenced earlier,
can yield new research questions about how digital storage complicates notions
of heated or cooled locations, as well as whether the definition of love letters
as hidden or private may matter differently if they are stored in a cloud or on
a flash drive. Everyday communication formats are changing, but it is unclear
whether the practices surrounding the storage of these communications will
change, too. My analysis offers a step toward these research paths, especially as
curatorial practices may relate to heteronormative expectations about gender,
intimacy, and family roles.

Notes

1 In many ways, this book perpetuates a heteronormative system by virtue of the use of
  words such as "he," "her," and "woman." The survey itself, while allowing for people
  to respond "other" to the question asking if they or their partners are male or female,
  treats gender, for the most part, as a binary (with only two categories) because the
  few people who selected "other" are omitted from most analyses for confidentiality
  purposes. In this chapter I approach a practice that is embedded in heteronormative cultural expectations by using some of its vocabulary. At the same I critique it
  in order to call attention to gender differences and inequalities that have real consequences in people's relationships today. This is a common dilemma in sociology
  and anthropology: using the categories at the same time the categories are redefined
  so inequalities are not reproduced or reified. Because the survey respondents overwhelmingly responded about heterosexual partnerships, it makes sense to talk about
  their experiences in this way. However, there is always a possibility that my survey
  wording or the topic of the survey itself was off-putting to people who would not
  identify themselves in a gender binary or as part of a heterosexual relationship with

one partner. I aim to acknowledge that not everyone may feel as if they fit into the rhetorical categories I use in my writing at the same time as I want to call attention to the significance of heteronormative romance in love letter curatorial practices as part of my project to uncover how powerful cultural expectations are.

2 If respondents list files or journals as the location of the love letter, they are excluded from both the paper and digital format categories, since it is not possible to discern from these words whether the letters are located in physical or digital locations.

3 Please see the Methodological Appendix for details about the sample characteristics in terms of sexual orientation. Important to note in this section is that the curatorial practices by gender are not differentiated by whether the romantic partnership was heterosexual or homosexual, but the majority of respondents referenced heterosexual partnerships.

4 For detailed elaboration of the survey findings as they relate to gender, see Janning and Christopherson (2015).

4

TIME MATTERS

Nostalgia, Preserving Love Letters, and the Social Construction of Time and Memory

A friend shared a story with me about how love letters in his family represented generational change. He described how, three years ago when his father passed away, he found his paternal grandfather's love letters to his grandmother in his dad's desk. He was sorting through his dad's possessions to decide what to keep and what to toss or give away. His grandfather was a traveling salesman around a hundred years ago, and the letters were all on hotel stationery that chronicled the various locations of his travels. My friend interpreted these letters in an interesting way. He noted that his father had been emotionally distant—someone who had a hard time expressing his feelings. But his grandfather, who he came to know anew through his letters, seemed to be quite expressive in written form. He wondered how it could be that a man in one generation could express his emotions so vividly while his son could not. The story mattered especially because this was a man who worked hard in his personal and professional life to defy expectations for men about being stoic or unemotional.

I wondered, in hearing this story, how this person's identity as a man was wrapped up in three generations of stories, some written and some shared in person (not to mention how expectations surrounding displays of masculinity have changed over time, calling to mind the heteronormative gender roles discussed in Chapter 3). I wondered how being away from someone and writing letters may or may not capture emotion the same way that speaking to them in person does. I wondered how something as individualized as identity is actually wrapped up in larger cultural norms about relationships, family roles, and what it means to be apart. As I situated this story in my larger questions about digital communication, I also wondered how age and generation may influence attitudes about whether the digital world is a good or bad thing when it comes to expressing romantic love. After all, to be a man today looks different than it did a hundred years ago.

Much of this chapter delves into how love letters tell a story about age, time, memory, and nostalgia—factors that also affect cultural constructions of what is perceived as a good relationship, good love affair, or good family. If people fall into what social historian Stephanie Coontz (2016) calls a "nostalgia trap" when they think about intimate relationships, their love letter curatorial practices may reveal a mismatch between what is idealized and what is really experienced. While this chapter delves into how time is socially constructed at a conceptual level and how age (as a proxy for time) matters in love letter curatorial practices, I also aim to point out how myths and idealized images of the past complicate expectations for current (and future) relationships. This is especially important when considering the benefits and drawbacks of digital communication technology in contemporary romantic relationships, and when considering how present understandings of past relationships may ignore important racial-ethnic, class, and gender inequalities that made the everyday lived experience of people vary.

In Chapter 2, I referenced the helpful work of Liz Stanley (2015) when I discussed the significance of time in the definition of love letter. In that discussion, it was apparent that the compression of time between exchanges caused by digitization of communication was evidence to some that the letter (and the love letter) is dying. To others, this was not true, and the digitization of letter-writing represented greater access to communicating these sentiments. Further, there have been many moments throughout history where increased communication speed was feared and then rendered normal and acceptable, suggesting that the belief that things are moving too fast is not unique to today's relationships.

So, time matters in debates surrounding the definition, and thus the classification as dying or not, of love letters. This is in large part due to the rapidity of digital communication. But how does time matter in terms of curatorial practices? Here, to center squarely on the importance of time, I share survey findings that shed light on how age and generation—important temporal factors of interest to social researchers—may impact love letter curatorial practices.

Age and the Digitization of Love Letters

Every time I see an older person asking a younger person for help in navigating a computer or smartphone in order to purchase a plane ticket, enter family members into a speed dial folder, decide whether to connect with a faraway family member in a social media platform, or look up a newspaper article, I am reminded of research that shows how age impacts the use and familiarity with digital communication formats, a pattern referred to as a generational digital divide. Specifically, new technologies have "an obsoleting impact on those without the proper skills and, with the speed at which the technology changes, it is very difficult—near impossible for some—to keep current" (McMullen 2016). I connect these findings to others that demonstrate a digital divide based on social group, knowing that access to digital ICT represents access to valuable resources such as the capacity to make online purchases, handle finances, feel safe, connect across geographic distance, and access timely news and information. In these instances, the older someone is, the less likely they are to show familiarity with new ICT—a digital divide based on how many years someone has lived on this planet. But I wonder whether this is a problem or not, given that some people are critical of how much digitization we've seen when it comes to interpersonal connectedness.

I am thus also reminded of how age may impact how people may view the benefits and drawbacks of an increase in the digitization of everyday practices, with older individuals expressing more concern than younger ones even at the same time they increasingly use new technologies such as FaceTime, texting, and social media sites to keep in touch with grandchildren and other family members who are geographically dispersed (McMullen 2016). In these instances, I always take a step back and wonder whether age may matter in ways that are not being measured or studied. In light of all of this, I wonder how age might impact curatorial practices surrounding digital and paper love letters. Below I share my survey findings that get at the intersections of age, attitudes about the use of technology in personal relationships, and curatorial practices surrounding love letters. I discuss how age matters in terms of: perception of societal patterns of digital and paper love letter writing; formats of romantic communication used at the beginning of romantic relationships that may shape the types of love letters saved; love letter curatorial practices; and the importance of the era in which the romance in question takes place.[1]

Perceptions about Technology

Given the aforementioned past research citing age differences in attitudes about the influx of digital technology into people's everyday lives, and as noted in previous chapters, I asked survey respondents whether handwritten letter, note, and card writing is fading in romantic relationships. My findings reveal that age does not seem to matter, with little variation between age groups (between 86 percent and 94 percent of each age group answered *yes*). Importantly, this question does not get at whether respondents see this change as positive or negative; it merely shows that the ability to notice the pattern of dwindling handwritten love letters spans all age groups. Nobody in any age group seems to be ignoring this pattern.

Communication Formats Used in Romantic Relationships

People who took my survey were asked to think of one romantic relationship where communication practices were salient. This relationship could be from the past or present, and it could be ongoing or have ended. In order to get at relationship communication behaviors, and thus set a context for people's love letter curatorial practices, I asked about communication practices at various stages of the relationship. Specifically, respondents noted the frequency of each of these types of written communication at both the beginning stages of the relationship and during the relationship: digital chats (Facebook messaging, Skype messaging, Gchat, IM, etc.);[2] emails; letters; texting; cards; and notes (including Post-its, memo boards, etc.). I then combined communication formats into digital written (digital chats, emails, and texting) and paper written (letters, cards, and notes) formats, and compared three age groups (18–34; 35–54; 55 and older) in terms of the frequency of love letter formats used at the beginning of relationships and frequency of those used during the relationship.

The central finding, which is not a surprise given the availability of certain technologies in the eras when survey respondents' relationships began, is that individuals aged 55 and older disproportionately used paper letters, cards, and notes in the early stages of their romantic relationships as compared to younger individuals (those aged 34 and younger), who disproportionately used digital formats such as email, digital messages, and texts. The one exception to this is paper notes, where there is not much discernible difference between age groups. While Post-Its were not invented until the late part of the twentieth century, it seems as if little paper notes given to a new romantic partner are as popular in today's relationships as they were in those that started decades ago, at least during the early stages of a relationship.

Once relationships have been going on for a while, couples may settle into behavioral patterns that look different from when they first started showing romantic interest. In addition to changes in relationship elements such as sexual intimacy, disclosure of personal information and habits, and connections to each other's social networks, communication habits can change. Frequencies of communication format used during relationships reveal interesting age group similarities and differences, some that resemble findings for early relationship stages, and some that do not.

Paper notes seem to transcend age in the middle of relationships, just as they do in the early stages. Paper cards are used more frequently during relationships for older individuals, and the use of paper letters is particularly low in frequency for the younger group. Digital chats and texting are more common during relationships for younger individuals, but emails seem to be most frequent during romantic relationships among the middle age group (ages 35–54).

Curatorial Practices

When I talk with people about their love letter storage practices, I pay attention to whether their age may impact how they talk about these practices. For instance, since I know that age impacts how settled someone may be in a location (older people are more settled), and that younger people today are moving around more than previous generations (Janning 2017), I suspect that older individuals may be more likely to save their paper love letters. They have more room and their home spaces are more permanent. Of course, given the results above concerning older individuals' greater frequency of using paper love letters in their relationships, they may simply have more paper letters to save. But decisions are not just based on whether someone can do something; they are based on whether a person may wish to do something. It may be the case that deciding to save love letters is impacted less by their format (digital or paper), and more by other factors, such as age.

And so, here I present my survey findings about whether age may impact love letter curatorial practices, with particular emphasis on saving and revisiting love letters. These behaviors include whether respondents save any mementos from the relationship, whether they save any paper or digital romantic communication specifically, and what type they look at most often out of the saved romantic communications (listed by types that include handwritten letters, notes, cards, emails, texts, and Facebook messages). In my analysis, the type of saved communication is collapsed into two categories: paper (letters, notes, and cards) and digital (emails, texts, and Facebook messages).

When asked whether they saved any physical or digital mementos that they would define as symbolic of the relationship, such as photographs, travel souvenirs, or gifts, the age of the respondent does not seem to matter. Just over three-fourths (76 percent) of all respondents save these things, with little variation between age groups (between 70 percent and 85 percent of each age group do this). When asked whether they save any pieces of communication from this relationship (the love letters), there are also no age group differences. Just over 88 percent of all respondents save these communication items, with little variation between age groups (between 85 percent and 94 percent of each age group do this). The people who responded to my survey, then, are similar across age groups in their high likelihood to save love letters or other relationship mementos.[3]

Respondents were asked to choose the type of communication they look at most often out of all types of saved communication pieces from the relationship; here there are age differences in terms of format saved (which, as is the case with the aforementioned findings about frequency of format use, may relate to era and availability of technology). Younger people—those 18–34 years old—are disproportionately likely to look at saved digital communication most often, and people aged 55 and older are disproportionately likely to look at saved paper communication most often. Importantly, though, despite age increasing the likelihood to look at a paper love letter frequently, a majority of people in *all* age groups (between two-thirds and nine-tenths) look at paper love letters the most often. Interestingly, when age groups are broken down into smaller clusters, the 35–44 year olds are more likely to save paper communication than both the 45–54 year olds and the 55–64 year olds. Thus, it is safe to say that younger people revisit digital love letters more than older people and older people revisit paper love letters more than younger people. But it is also safe to say that paper love letters are preferred as the format of letter to be revisited across all age groups. It is *not* safe to say that revisiting paper love letters *only* happens among older individuals. In fact, among the people between the ages of 35 and 64, the younger the person, the more likely she or he is to revisit a paper love letter. This is all to say that saving and revisiting paper love letters is not reserved for older individuals, even if it is labeled by some as old-fashioned. This means that factors beyond age and availability of communication format may impact love letter curatorial practices. These factors—larger cultural, political, and economic ones—are elaborated in Chapter 5.

The Importance of Era

As mentioned earlier, the era in which the relationships occurred or are occurring necessarily affects the frequency of use of the communication format. For example, my parents did not save their text messages from their dating years because texting was not around in the early 1960s. However, recall that people who took the survey could use any romantic relationship from their past or present, so I could not assume that age would perfectly overlap with era. Add to this the fact that people get involved in romance at different ages and life stages, and it is necessary to isolate the impacts of age from the impacts of era.

Because the era in which the communication occurred depends on available communication technology, I asked when the salient relationship began and then collapsed these responses into three eras that coincide with significant technological communication changes: before 1996 (when the World Wide Web was commercialized and more readily accessible by the public); between 1996 and 2007 (when email, texting, and the internet were growing in use); and between 2008 and 2013 (just after the iPhone was released, and just after Facebook became public and started to see a significant increase in users). I then studied whether there was a link between age group and the era in which the relationship started in order to see if the time when a relationship started was more recent for younger respondents and farther in the past for older respondents. I also examined the connection between era and type of saved communication most often looked at to see how closely aligned the type of communication was with the available technology at the time.

Without going into the statistical significance test nitty-gritty, and despite acknowledging that there may be variation in relationship start times within age groups, suffice it to say that there is a very strong relationship between age and the era in which the relationship respondents referred to in the survey began. This makes sense, given typical ages of romantic relationship forma- tion (and given that survey respondents who are 18–24 would have been babies or toddlers before 1996, and could therefore not have chosen that timeframe for the beginning of the romantic relationship). Importantly, though, not all younger respondents started their relationship between 2008 and 2013, and not all older respondents started theirs before 1996.

Connecting era and communication format also makes sense. After collaps- ing the time when relationships started into three eras, it is clear that digital saved communication items were disproportionately looked at often in more recent years, while paper saved communication items were disproportionately looked at often before 1996.

Why include age if era and availability wipes out most of the impact on love letter curatorial practices? I include this because past experience shapes atti- tudes and current behaviors. So, while I offer these survey findings with a grain of salt given the overlap of age with era and availability of technology, I still think it's worth examining. In particular, the fact that age doesn't seem to matter much in people's likelihood to save and revisit paper love letters more than digital ones suggests that the availability of a communication format may not entirely explain people's preference for it. This matters because, as men- tioned earlier, people of all ages have noticed that paper handwritten love let- ter writing is fading away. So, there seems to be a mismatch between the objects people prefer to save and revisit and whether those objects remain in frequent use. The mismatch between what is preferred and what is actually happening crosses all age groups, a finding that hints at the allure of remembering the past in ways that may capture ideals (or, as discussed in Chapter 2, the "Sunday self") more than realities (the everyday mundane self).

Saving Love Letters and the Allure of Nostalgia

In this chapter, I use the concept of time in different ways. I refer back to the debate introduced in Chapter 2 about how taking time to handwrite a love letter, desirable to some and unnecessary to others, is an important part of the deliberation in defining something as a love letter or not. In that sense, time is about the letter writing, receiving, and interpreting processes. I also use time to refer to generational change in terms of how letters are written and how they are viewed once they are found, as told in the opening story and as shown in my presentation of survey findings. Finally, I refer to time in terms of era and generation, which are used to create boundaries around cohorts of individu- als in order to show how their social experiences may be similar because they experienced them around the same time. Time is socially constructed, which means that, rather than looking at a year or a decade in absolute terms, social researchers point out how people decide whether and how time matters in their individual and collective stories. Put another way, I may not remember what happened at 11:25 a.m. yesterday (other than it preceded my lunch), but I think it's interesting to study why remembering this may matter, why people use certain notations to demarcate time, and how these demarcations shape the ways people make meaning in their everyday lives.

I propose that time plays a role in how people conceptualize, categorize, and portray cultural values, both collectively and individually. Events are signified temporally. Things are preserved in the hopes that they'll withstand the test of time. Time periods and events, accurate or not, are referenced in order to assess what's going on now. This hearkens to Stanley's (2015) cautionary words about the problems of being ahistorical. Sometimes the past is misremembered or people use misinformation about "how things used to be" in order to critique how things are today. This reference to the past in order to judge the present has real consequences. As framed in sociology's *Thomas theorem*, by defining situations as real, they have real consequences. How someone remembers the past shapes how they view their present and their futures.

In fact, as Norwegian anthropologist Marianne Gullestad (2004) has said about research on childhood memories, sometimes how the past is viewed has more impact on present identities, roles, and experiences than what actually happened. At an individual level, then, time matters in how people construct memories. Mementos such as love letters are mile markers on memory paths. The commemoration of a past moment in a relationship can shape how people think of themselves today. At a cultural level, memory matters in how people think of the past, too. The collective remembering of an event or historic moment, accurate or inaccurate, affects how people make decisions and assess what's currently happening. Sometimes this remembering is materialized in monuments, memorials, or other physical renditions of a historic moment. A walk along the National Mall in Washington, D.C., shows this effectively, since the desired effect on viewers of Maya Lin's low and humble Vietnam Memorial is quite different from that of the majestic fountain-filled World War II Memorial.

Sometimes when I've given talks about love letters as cultural objects, audience members will discuss with me afterward about their wistful wish that it'd be great to go back in time and bring back the importance of love letters. They grieve the decline in pen-and-paper love letter writing, sending, reading, and saving. Their lamentation usually goes something like this: "Wouldn't it be great if people still wrote love letters like they used to? People sure aren't as romantic as they used to be. Let me tell you a story about my parents . . ." What matters here is not whether people *actually* used to write more pen-and-paper love letters in the past as compared to the present (which they probably did), but rather what someone's perception of the past does to affect their judgment of present circumstances (which can have just as much impact as what really happened).

Another cultural object that illustrates this process is e-readers. Kindles, Nooks, and books downloaded to smartphones have made reading books more portable and accessible than ever before. But despite the growth of e-readers in the marketplace, some people cling to their preference for holding a hard copy of a book in their hands. This preference may be defined in terms of the feeling of closeness to an author that comes from flipping back to see cover art and the author's name. Or it may be defined in terms of the bodily experience of feeling the weight of a book in their hands. Or it may relate to being able to check progress using page numbers or the thickness of the pages that come before and after their stopping point instead of percentage of the book read or number of minutes left to finish the chapter if the same reading pace is maintained.

What is a book? For those who require a cover, the weight of a board and paper object in their hand, and being able to see page numbers, only a printed book counts. In fact, it is these physical traits that matter in how they read, understand, and evaluate the story. For others who define a book more based on its intent to convey a story, the format may matter less. Whether progress is measured by percentages or page numbers, people who prefer e-readers might say that the digital version is just as real as the physical one. To cite the Thomas theorem again, if a person defines an e-reader as real, that person will experience a book as an actual book. If a person does not define an e-reader as real, the consequence may be an inability to define the experience as book reading.

What's so alluring about the past? What are people grieving, whether it's about pen-and-paper love letters or printed copies of books that have covers and page numbers? The *fear* of losing something important can play a big role in how people perceive what's going on in the world. People cling nostalgically to the past because change is hard work. As McMullen (2016) points out, "things that we have to learn are difficult, while things we grow up with are just part of the environment—something we always knew." But nostalgia is also about judgments about that change and a clinging to bits and pieces of the past in contrast to viewing today's whole story. The allure of nostalgia is that it makes people feel grounded in moments of rapid change where deep-seated fears make them and their way of doing things feel threatened. That's the fodder of every story about generation gaps that usually begins with something like "Kids these days . . ." But standing on shaky ground filled with misinformation or selected bits of history that are only called forth in order to protect a desire to do it in one certain way, in the long run, is a risky spot to stand on.

As my survey findings reveal, age seems to matter less than other factors in people's desire to save paper love letters over digital ones. While the likelihood to save a paper letter is impacted by whether paper love letters are sent in the first place (something more common decades ago than today), perhaps the preference is about nostalgia, or keeping alive the memory of a perceived simpler time.[4] If that is the case, it is interesting that age doesn't seem to matter in terms of the allure of nostalgia. But perhaps the way in which nostalgia operates varies by age. Perhaps older people may want to go back to a simpler time because they lived it, whereas younger people may want to go back to a simpler time because they've been socialized to view it as romantic (and simpler than the present) by parents or in popular culture. The outcome is the same: romanticizing the past as a simpler time when an idealized form of romantic love is perceived to have been actualized. In this scenario, the cultural processes at play that lead to the outcome vary by age, but people of any age may fall victim to the nostalgia trap. And, of course, there is nothing simple about what happened in the past. People just like remembering it that way.

Playing With Time: Imagined Future Kinship Nostalgia

Why do I save my son's school papers? Is it because I saved mine and enjoy looking at them occasionally so I can reminisce about childhood and feel proud of my stellar organization system, thus projecting my desires onto his? Is it because I don't think he's equipped to know whether he'll want them later, and I don't want him to regret not saving them, thus making me see him as a person who is not fully capable of making a decision and therefore necessitating me to take care of his future emotional well-being? Is it because we did this in my family, so I'm just following through on how I was socialized? Is it because prescribed idealized parenting roles suggest that parents (especially mothers, as discussed in Chapter 2) must dedicate time and energy to raising children, which includes keeping material reminders of their success as parents? I've highlighted the importance of looking back in time, correctly or incorrectly, and seeing how that historical view shapes perceptions of the present. I also think it's good to look to the future, to how today's experiences, especially love letter curatorial practices, demonstrate a future orientation that is about the preservation of relationships beyond the relationship itself, including the preservation of a family's story and of collective values. Saving a child's school papers is not just

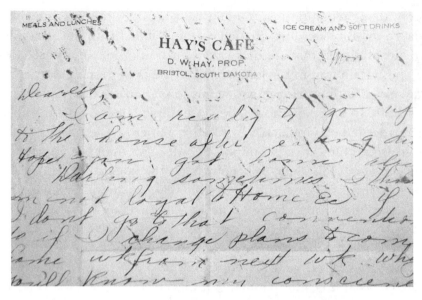

Figure 4.1 **Saved Letter From Grandparent** (Source: Author Photograph)

about one parent and child. It is about the longevity of generational connections and the culture of parenthood and childhood (and education practices), too. Saving love letters is not just about preserving one romance. It is about the cultural preservation of romantic love, too.

I've done research, mentioned in earlier places in this book, on parents' curatorial practices surrounding family photographs, both digital and paper. As with family photos, when it comes to saving, organizing, and sharing, sometimes it's not about the present. Sometimes it's about an imagined future where, once someone is gone, the hope is for their story is told in a favorable light. It is in individuals' interest today to curate their story for an imagined future audience who, they hope, will judge them positively. Today's children are imagined as tomorrow's adults. Curatorial practices may be the way they are because people want the next generation to look at an organized story from the past, and to share a favorable rendition of a contemporary love story for decades to come.

Conclusion

I love that my mom showed me her parents' love letters—letters written at kitchen and small-town café tables in Minnesota and South Dakota in the 1930s. She even let me keep a couple of them, which I have re-read dozens of times and which I store in a box with other family mementos that came into being at a time before I was born. I love the loopy cursive handwriting written in black ink (see Figure 4.1). I love that they are multiple pages of folded thin paper. I love that the envelopes did not need zip codes or precise addresses. I love the stories my grandparents shared with each other in writing about their time living apart: what was going on at the farm, how his schooling was going, how her work was going, and, as is typical in Midwest conversations even today, what the weather had been like. My grandparents had to live apart for some of the early stages of their marriage. My grandma was a teacher, and, as the story goes in my family, she would not have been allowed to keep her job if she was married. So, she lived in a boarding house in a different town. They kept their marriage a secret and wrote letters to each other for several months while they were both going about their days.

These letters, to me, brought to life my grandparents' love story. They also brought to life all of the stuff I had learned in college and graduate school about 20th-century changing gender roles, work, and rural life for families that were not well-to-do. They brought to life the inequalities that centered around gender and geographic location, in particular. If I hadn't been studying these things when I read the letters, I may have been tempted to gloss over the inequalities that structured my grandparents' intimate lives. I knew that it wasn't a simpler time, at the same time I looked at the letter as a quaint symbol of that simple time. I saw my grandmother's personality represented in black ink on the stationery she found at a café in the town where she taught in rural South Dakota, but I could not just look at it as a quaint memento. The letters, by the way, also had some steamy parts that made me appreciate the desire my grandparents had for each other, especially expressed when they couldn't live together or let their workplaces know they were married. As this story shows, even I, as someone enmeshed in academic thinking and writing about the problematic allure of nostalgia, especially as it connected to my own family's story, wanted to view these letters as ideal mementos of a simpler time with fewer complexities than today's hectic world brings to me and my family.

I appreciated that my friend, mentioned in the introduction of this chapter, shared his story about his grandpa's letters with me. At the end of our conversation, he offered to let me see the letters (this is a common offer for a researcher who studies love letters). My response was that, at least for my research, I was more interested in what people do with the letters and how they think about them than I was interested in their content. To him, though, the content was everything. It wasn't just that there was a stack of letters in his dad's desk; it was that the letters had particular words written in a particular way to a particular person—all particularities that he had determined represented how his grandfather and father expressed emotion differently; all particularities that made him ponder his own identity, an identity wrapped up in three generations of men figuring out ways to communicate their feelings to someone else. One wonders, though, what bits and pieces of a family story may be lost when the third generation has only paper letters to tell the history.

At the level of family, the things people already know about their loved ones shape their impression of the artifacts left behind. This was true in my own family, when I interpreted, correctly or incorrectly, my maternal grandparents' love letters as confirmation of early signs of relationship problems. To clarify, their marriage, like many, was flawed and filled with conflict, especially later in their lives. Because I knew that, I read those 1930s letters looking for signs of early problems. Even in my quest for disrupting the nostalgic view of family life and for ensuring that every family story needs to be embedded in understanding about larger social and historical forces that I had been learning about in graduate school, I looked at the letters with a selective lens. No amount of academic understanding of the structural and cultural influences on their relationship would have mattered in my understanding of their story, because I was looking for the story to be told in a certain way and I wanted their story to help me think about my own family life. Beyond family stories, cultural values shape not only how people view the letters they find in their parents' desks, but also how they view their own identity as it relates to romance, love, and happiness.

Of course the content of love letters matters. But what is worth pointing out here is that someone's perception of what the content means and why the content matters is worth analyzing. Of course it is unlikely that my grandma wrote her letter with me in mind as an imagined audience. But I wonder if people think about how their love letter may be saved as they write it. I wonder if imagined future curatorial practices shape the decision to write, how to write, and what format to use to write. I wonder, most vividly, whether a nostalgic view of love letter writing shapes how someone may imagine their future love story to be told.

Notes

1 It is important to note that people aged 18 to 24 were disproportionately represented among the survey respondents because one of the primary methods of distributing the survey link was via a college student listserv. However, there are enough people in each age category to show group similarities and differences. I therefore compare age groups based on within-category proportions rather than raw numbers, which matters for any statistical significance tests I conducted. When I discuss group differences or variations, I only note variation if the age group differences were found via Chi-square (for questions with categorical responses such as yes/no) and Analysis of Variance (for questions where age group means are compared) statistical significance tests.

2 Captured Snapchats were included in the survey as an option for saved communication, but the number of people who selected that in this particular analysis about age differences was very small, so they are excluded here.

3 Do nine out of ten people in the larger population save letters, or is this practice something that is found only in these survey results? While it might be tempting to suggest that these high numbers are also represented in the larger population, including those who did not take my survey, it is important to note that *people who are willing to take a survey about love letters may be disproportionately likely to save them in the first place.* This is why it is okay to focus on group differences within the sample, especially if the groups are large enough, but it would be inappropriate to assume the findings for the entire sample are representative of the larger population. Further research conducted with more attention to representative sampling is warranted.

4 Importantly, in contrast to this discussion about the allure of nostalgia and misre-membering the past only in a positive light, at times people save love letters in order to maintain a realistic view of the past. This can happen when someone saves a letter that may call forth negative memories or display immaturity in order to remind themselves of how far they've come in life. Present identities are shaped by how people think of past experiences. Sometimes people want to forget about them; sometimes they want to call them forth. In either case, keeping a letter to bring forth negative memories is still an example of using a perception of the past to shape how people think of themselves today.

5

LOVE LETTERS AS BOTH INDIVIDUAL AND COLLECTIVE

The Public Significance of Private Communications

In a January 2018 story featured on National Public Radio, one couple's romance—Elizabeth from Michigan and David from Wales—was chronicled in terms of their use of apps such as Omegle and Snapchat to meet and then communicate across geographic distance. The digital messages they exchanged, some of which they saved, formed the "building blocks of their relationship" (Roman 2018). Even without saving every message, they were able to recount the life-changing message where David declared how much he liked Elizabeth, which led to a decision that they'd meet in person. Nervous about meeting, Elizabeth posted her concerns and declared her love for "someone from across the Atlantic" on an online long-distance relationship subreddit, "a forum on Reddit where people can share their relationship experiences." David found the post, asked her about it, and told her he loved her, too. They're still together. They define digital platforms such as Snapchat and iMessage as "the love letters of our time period," and the article included numerous screenshots of their texting messages that they had preserved.

There are several interesting features of this story. First, the pair define digital communications as love letters. Second, they laud the existence of digital communication technology because it is what allowed them to meet, stay connected across geographic distance, and have frequent accessible reminders of their love. Third, their story was shared publicly on Reddit and they allowed their saved text exchange screenshots to be shared as part of this public radio story. These features show that Elizabeth and David include digital messages in their definition of love letters, they feel as if they have maintained personalization and connectedness across geographic distance, and they appreciate the speed with which they can communicate. Their story was preserved digitally in accessible places, and they are not afraid of sharing their story publicly. They do not fear losing love letters; rather, they fear that without digital communication, they'd have lost their love (or, to be precise, they'd never have found it in the first place). This begs the question raised by Stanley (2015: 242): "And is the letter actually so important, or are there more fundamental aspects of epistolarity that might even be enhanced by current new developments?" In other words, does the digitization of love letters allow for them to be enhanced and made increasingly present, or is there something about the paper-and-pen version—the format my survey respondents prefer to save at the same time they say is fading away—that makes digital developments threaten the longevity of love letters?

Another story illustrates these questions further. On the one-year anniversary of his wife Catherine's death from ovarian cancer, Hyong Yi honored her memory by handing out 100 paper letters to strangers, letters that were back-and-forth exchanges between the two of them that spanned their entire relationship. He did not intend for this effort to instigate a mass movement, but the presence of #100LoveNotes on Twitter and the publicity surrounding his efforts show that people are invested in a movement to share reminders with people they love of how much they mean. In the news story highlighting his project with the title "The Way This Husband Is Honoring His Late Wife Is a Beautiful Testament to the Power of Love," the writer advised: "Take a moment to reflect on those around you who make you smile and bring joy to your world—and tell them just how much they mean to you. It can be a simple text, a Facebook post, a phone call, or even an old-school letter" (Shoaff 2015). What is intriguing about this story is twofold: first, the initial love letters were paper with personalized handwritten lettering, but the project has moved to a digital platform to allow for the pace of sharing to accelerate; second, Hyong Yi wanted to preserve the memory of his wife by sharing their private moments publicly. The letters were removed from their storage place and dispersed into the hands of strangers, whose next steps were unknown to Hyong Yi. Maybe they gave them to someone else. Maybe they put them in a drawer. Maybe they threw them away. This story, captured in the phrase in the title "The Power of Love," shows how powerful the cultural value of romantic love is. Whether it is readers of the story or participants in the #100LoveNotes Project, people beyond the couple are invested in someone else's love. They are invested at an even greater level because the letters, shared initially on paper, have becoming digitized online, widening the audience. While the project's goal was to honor a woman and make people be more intentional about acknowledging those whom they love, the project (along with the story about it) carries sociological significance: it lifts up the status of romantic love as something that is to be achieved, a private exchange that symbolizes a collective value.

The interplay between the individual love letter as exchanged in private between partners and the collective project of preserving cultural values associated with romantic love is the focus of this concluding chapter. I overview other scholars' writings that situate how this interplay matters for love letters. I propose a framework that can be used to understand how people view the longevity of love letters in the digital age. I suggest that the assessment of digital communication as a threat or enhancement to the existence of love letters (and, by extension, romantic love) depends on different definitional elements of love letter that align with the fear of losing different cultural values. I conclude with how any consideration of love letters as symbols of how culture works needs to invoke the significance of privilege, thus making the study of "stuff" relevant in larger questions of power and inequality in society.

Choice, Constraint, and Culture: Love Letters as (a) Social Matter

Why does any of this matter? Why do private lives matter in the public sphere? Why do objects that are tucked in people's basements or smartphones tell an important story about social relations? By studying digital and paper love letters and what people do with them, five important connections between seemingly individual actions and broader cultural values are revealed.

First, whether saved in a storage box in a basement or organized in a cleverly disguised folder on a laptop, love letters are part of the material world. They are objects. What counts as an object can range from a digital photo or screenshot of a paper letter to the paper letter itself, as complicated in the image of me taking a picture of my grandma's letter in Figure 5.1 (where I'm digitally preserving the digital preservation of a physical object!). The material world tells an interesting story about the social world, or, as some anthropologists and sociologists may say, *material culture* can symbolize *non-material culture*, especially in terms of identities, relationships, and statuses within a larger system of groups. As Jenkins (2017) notes, one "cannot close the doors of a home around a romantic relationship and imagine that this prevents society from intruding" (131). Love letters as material culture, and curatorial practices including whether and where people store them, how people interact with them, and what format they're in "play a big role in people's accomplishment of social identities that are much larger than any individual preference. Objects (and the spaces those objects occupy), thus, tell us about broader social issues in the form of groups, status, and cultural context" (Janning 2017: 11). Certainly if the content of a love letter matters to someone to whom it is sent, they may be more likely to save that message as a reminder of devotion or admiration.[1] Letters are the textual renditions of identities and interactional processes. Saving and looking at them involves the processes of self-definition, discovery, and disclosure, and the social processes and audiences that surround these. To understand romantic relationships, it is helpful to look beyond the content of love letters and toward the curatorial practices surrounding them, especially since those

Figure 5.1 **A Digital Photo of a Digital Photo of a Paper Love Letter** (Source: Author Photograph)

practices and the relationships themselves are impacted by changes in communication technologies.

Second, everyday practices that seem private actually demonstrate an investment in a larger system. As evidenced by the warm reception of the #100LoveNotes project, people today are invested in a system of love letter writing and exchange, albeit one that is undergoing rapid technological change. People are invested the same way high school students are invested in the dating system (Milner 2016): they pay attention to others' stories that affirm a belief in the values associated with romantic love. They even exchange material symbols that serve as reminders of the system's existence and strength. For high school students, the visual representation of pinning, wearing a letter jacket, changing a Facebook status, or donning a ring not only reminds them of their participation in a system (if they're in a couple donning the symbols), these practices also display the system to an audience whose investment perpetuates the system. Similarly, even if someone does not write, send, or have love letters, their belief that love letters are material objects that help people maintain romance shows how much they support the system of romantic love. Individual curatorial practices and beliefs that love letters ought to be curated demonstrate investment in the system of romantic love.

Third, there are countless cultural (and popular cultural) locations that perpetuate collective beliefs about romantic love, which influence individual practices. As philosopher Carrie Jenkins (2017) says, "romantic love is not straightforwardly an individual or private matter." Literal elements in society shape a definition of love played out in romantic comedies, love poems, and song lyrics. Figurative elements, or "the expectations for a 'normal relationship' that we have been absorbing ever since we joined in playground renditions of the K-I-S-S-I-N-G rhyme" (132), do this, too. Love letters as objects can be both literal and figurative elements of romantic love: they symbolize a collective definition of romantic love in the telling of individual love stories at the same time the individual curatorial practices surrounding them show what cultural values matter surrounding romantic love. Similarly, as people read and say the kissing rhyme to define love, their everyday practices (such as kissing a romantic partner) demonstrate whether they meet the expectations for kissing as a way to show love.

Fourth, the classification of love letters and their content requires a collective definition of romance at the same time the content is crafted individually. While an individually written love letter may contain inside jokes, personalized font/ink color, or reference to intimate memories that only the recipient understands, it also may be classified as a love letter because it contains reference to collectively defined "subjects such as courtship, marriage, emotional attachment, and intimate disclosure of the self and sexual feelings" (Teo 2005: 344) that situate the letter in the category of love letter. However, as I discussed in Chapter 2, there can often be reference to other matters in a communication that is labeled "love letter," (e.g., a grocery list), and there can be communications that are defined by their sender and recipient as a love letter that have no mention of these topics, thus counting for these people as love letters anyway. What matters here is twofold: first, there is choice about which content to include but the potential ingredients for that content are collectively defined; and second, as with any form of deviance from a norm, the claim that a letter strays from normal and is defined as a love letter *anyway* reinforces the existence of a collective definition in the first place. In all of the vast research on love letters, there is evidence that there have been enough patterns in place in the last century or two to show some standardization, especially in terms of content and epistolary intent. In addition, as with many emotions, the feeling of being the only one who has ever been in love has to be situated in the larger context of a collective belief that feeling unique is part of the love experience.

Finally, love letters call to mind the importance of group differences and inequalities that allow or constrain someone's ability to write, read, save, store, access, revisit, toss, or organize them. As the previous paragraphs have outlined, individual love letter curatorial practices fit into a larger system that is organized and defined by larger forces than any individual lover. The system and practices within it are also politicized and must be situated in terms of powerful values associated with today's economic conditions. Illouz (1997) argues that this is because the practices reinforce a *hegemonic* (most dominant) romantic love ideal within a literate capitalist society. She recounts her middle- and upper-class interview respondents' stories of past relationships, noting that they "engaged in the process of weighing positive or negative attributes in order to maximize satisfaction" (218). Guiding all of these stories is a framing of psychological needs motivated by self-interest. Further, "[t]he rational assessment of personal attributes has clear economic overtones, for it is used to shop in the supermarket of relationships for the partner most likely to satisfy personal preferences. . . [this approach is] culturally and historically specific" (219). She continues by saying that the "people most likely to use the rational conceptions of love promoted by popular culture are upwardly mobile and concerned about maintaining and maximizing their social status" (220). Illouz's more affluent interviewees had detailed lists of desirable attributes in partners, although they resisted saying they'd choose partners based on social and economic assets, treating "money and sentiment as distinct categories" (221). Later, Illouz notes the significance of communication as it relates to class-based romantic love practices. Specifically, she tells the story of a working-class man's attempts to seem desirable to a more educated woman. His story highlights that "verbal communication was one of the important avenues, if not the most important, through which his educational inadequacy or incompatibility was made visible" (234). Romantic love in its ideal state, then, is

> talkative love . . . verbal intimacy. . . [U]nlike the seventeenth-century aristocratic French ideal of sharing tastes through the refined art of conversation, the aim of contemporary communication is not to engage in sophisticated seduction games, in which the self is as often veiled as it is disclosed, but rather to expose and express, as authentically as possible, one's inner thoughts and self.
>
> (234)

Communication, then, is supposed to occur in couples in order to maintain intimacy and emotional self-disclosure. This ideal is present in self-help relationship books, in popular media, and in research findings by Illouz and others.

How one fares with language varies by economic and education level—the linguistic form of what French social theorist Pierre Bourdieu named *cultural capital*. Thus, even the participation in love letter writing itself is situated in larger political and economic inequalities.

While much of this discussion centers on verbal face-to-face communication, elements of class-based self-presentation and idealization of certain ways of communicating can be present in written communication, too. Illouz, thus, would likely situate love letter writing, saving, and storing in a larger system that is guided by a value of self-interest, group inequalities, economic exchange in a marketplace where emotional connections are commodified, and a strong belief, perhaps paradoxically, that experiences and expressions of romantic love are supposed to be separate from the presumed emotionless work attached to money and selfishness.

Love and Fear in the Digital World: The Cultural Work of Emotions

In the introduction to this chapter, Elizabeth and David's story highlights the fact that, while they fear losing closeness by virtue of their vast geographic distance, they believe digital communication gives them an opportunity to share and maintain their love. Their personalized messages, rapid pace of exchange, ability to preserve the messages to revisit, willingness to share their relationship beyond their private world, and the storage of the frequent messages in a place they can easily reach all add up to their ability to define their digital exchanges as love letters. But not everyone may agree that this form of romantic communication constitutes love letters. Here I examine what parts of the definition of love letter may matter in someone's likelihood to fear that they're going away, especially in light of the influence of digital communication technology into romance.

To save something means to prevent it from being lost, as in "her life was saved; her life was lost." In the #100LoveNotes story at the beginning of this chapter, a woman lost her life but her husband sought to breathe new life into the loss by sharing their romantic love story beyond the private walls of their home. Clearly, to recover something that is lost, even if symbolically, is a desirable social action that adheres to larger cultural values about love and loss. So, what would be lost if love letters were lost? What are people afraid to lose? And, how is the question of loss wrapped up in changing technologies that have diversified the formats love letters have taken? These questions are tackled in this section, where I lay out how all of the topics that have been discussed in this book come together to form a new way to organize how to think about love letters in a digital world—namely, in terms of *what people fear they'd lose.*

Are people really afraid of losing love in a digital world? What separates people who answer yes to this question from those who answer no? In this book,
I have shared ideas and interpretations about how love letter format (digital or
paper) and people's love letter curatorial practices add to an existing understanding of how culture works. Much of the discussion about the impact of digital formats on love letters centers around fear and loss, including, for example,
fear that love letters will be rendered less personal with the use of emojis, or
that a "box" of love letters will be more easily lost if they're stored in a digital
cloud. These fears surfaced in my survey findings, they are theorized in other
scholars' research and writing, and they were animated in stories that people
have shared with me as I have continued the conversation about love letters
with others. But what do these fears mean?

In Chapter 1, I noted that romantic love, as an emotion, needs to be examined as a cultural concept. This means that it is necessary to take a step back and
ask which social processes have played a role in defining what is thought of as
a highly individual and psychological (and sometimes chemical) thing. Values
and belief systems that are collectively shared (and that may vary from large
group to large group) need to be examined because they influence individual ideas about seemingly personal matters. In fact, these larger cultural values
help to shape what is defined as personal in the first place. Culture is dynamic,
with both subtle and drastic changes that occur over time.

In essence, romantic love and its representation in the curatorial practices
surrounding love letters is situated in a spectrum from constraint to choice.
People succumb to collective beliefs about romantic love at the same time they
shape its presence in their own lives; they are shaped by social forces as they
choose their own curatorial practices. So, while someone may feel innovative
for typing in old text message exchanges in a spreadsheet or for decorating a
box that is stored in a closet, they are still constrained by the technologies that
allow or disallow spreadsheet-making and by the built environment and marketplace that allow or disallow the existence of closets or doors or decorating
supplies for storage boxes.

As scary as it may sound because it may shock people into challenging long-
held myths, understanding how romantic love is cultural is tremendously
important. I recommend doing the same thing for any emotion, including fear.
Emotions need to be examined not just as individual experiences, but as cultural ones, since they vary depending on cultural context, there are patterns
of emotional experience and display within groups, the experience and display of emotions are affected by expectations operating in culture, and often
adhering to normal ways of experiencing emotions leads to greater well-being.
This is because someone who feels as if they are a good and normal person

will be more likely to feel a sense of belonging and satisfaction than someone who doesn't (Mesquita et al. 2016). But what happens when the culturally prescribed emotion seems counter to well-being? What happens if the experienced emotion is fear? If, as Mesquita et al. (2016) assert, "individuals seek out situations that foster emotions that are useful to culturally central tasks" (32), then what tasks are accomplished if someone feels fear?

Fear as an emotion has been examined sociologically from a *macrosociological* (large group or societal) level, as in studies about the culture of fear surrounding crime and other social problems and in studies about risk and fear of disasters (e.g., Glassner 1999; Ferudi 1997). Fear has also been studied as an emotion from a *microsociological* (individual or interactional) level, as in studies about how interpersonal relationship problems may instill fear or in how fear as an emotion demonstrates how the self is constructed in different social settings (e.g., Hochschild 2003). In all of these cases, there is a common belief that "even the most seemingly straightforward situation of fearfulness is heavily mediated through the physical, psychological, culture and social environments in which it is located" (Tudor 2003: 240). Fear exists within an "emotional climate"—a collective "cultural matrix" that has patterns of behavior, attitudes, and beliefs that affect people's individual experiences. The experiences are sometimes a lot of work. Fear, like love, requires emotional labor, especially if it involves challenging "an individual or a group's idea or fantasy of what their interest—and what the public interest—actually is" (du Plessis and Sørensen 2017: 182). Thus, it may be that if people perceive that their beliefs and practices surrounding romantic love are challenged, they will fear the loss of symbols, like love letters, that represent their beliefs. Today's digital society represents an interesting crossroads when it comes to the fear of loss of symbols, in light of rapid technological change that forces the question of what the definitions of symbols are. What is a real love letter: a piece of paper or text held in a digital cloud? Someone's answer to this may depend on how much they fear the loss of current cultural conceptions of romantic love.

A New Conceptual Framework

Should people be afraid of a futuristic AI robot monster who becomes a substitute lover and dehumanizes the world to the point where people are incapable of love, clear communication, and writing a cursive letter *L*? The answer to that, at least as I think about it sociologically and anthropologically, depends on what part of the monster is seen and viewed as significant.

I propose a framework that defines the fears surrounding the digitization of love letters, with the central fear defined as losing cultural values associated with contemporary Western conceptions of romantic love.[2] As with love, outlining fear necessitates understanding the cultural processes that create, support, or challenge individual attitudes and behaviors surrounding whether and how people save, store, re-read, organize, stumble upon, lose, burn, or hide digital and paper love letters. If people fear the loss of romantic love, it follows that preserving romantic love symbolically, as with love letters, may serve to lessen that fear. Romance, in this case, is preserved as evidenced by people still writing love letters to each other. At an individual level, someone may decide to write love letters because they fear that not doing so would lessen romance in their life. But it's more complex than referencing an overarching central fear about the loss of love. There are *different fears about the loss of different cultural values that are associated with different elements of a love letter's definition.* Thus, what is feared to be lost may differ depending on which part of the definition of love letter people think matters. The fear can be connected to any or all of the following elements that make up a definition of love letter—personalization, pace, preservation, privacy, and place. For someone who believes a love letter must be personalized, slow in the amount of time it takes to write and respond, savable, private, and stored in a secure spot, their definition of love letter contains all five of these definitional elements.

The elements of the definition of love letter, if perceived as being absent by virtue of love letters being digitized, connect to different fears about the loss of different powerful cultural values. Put another way, what people are afraid of in terms of love letters fading away as a result of increased digitization depends on which parts of the definition of love letter they think matters. Fear about the potential for losing love letters is really about what powerful cultural values would be lost if certain elements of love letters went away. Here I propose and explain a conceptual model that encapsulates love letter definitional elements, along with the cultural values that would be feared to be lost if that element was missing (Figure 5.2).

Personalization

Imagine someone defining a love letter as something that has to show the unique traits of the person sending it. This could be exemplified by a special writing style or the inclusion of an inside joke in the letter. In terms of curatorial practices, maybe the storage container is decorated or labeled differently than other containers, or maybe the storage location for the most important lover's letters is different from the storage location for other lovers' letters.

If love letters are digitized, some may fear that personalization may be lessened because there are fewer elements of the "person" present. This concern was raised in Chapter 2, when I outlined Stanley's (2015) discussion of how people fear a digital *de*personalization if a letter doesn't have the stylistic

Love Letter Element	Cultural Value Feared Lost as a Result of Love Letter Digitization
Personalization	Individualization and self-interest
Pace	Time-taking, patience, and depth in a hectic world
Preservation	Individual and collective legacy and longevity
Privacy	Preservation of the public-private boundary and the intimacy of the monogamous dyad
Place	Storage of cherished possessions in secure, important, and accessible locations

Figure 5.2 **Digitization of Love Letters and the Fear of Losing Cultural Values: A Conceptual Framework**

conventions made possible with pen and paper. If someone's handwriting isn't in a letter, then it doesn't count as personalized, thus making this definitional element of love letter absent. No longer would the letter or collection of letters be individualized. No longer would love letters be about the unique traits of the partners (both the sender and the receiver), traits that are considered by the lover in their desire to fulfill their own self-interest, as Hochschild (2003) suggests. To depersonalize either the love letter itself or the practices surrounding the writing, sending, or keeping it is to threaten the value of individualization and self-interest. Fearing the loss of personalization that may be due to the digitization of love letters serves to reinforce powerful cultural values such as individualization and self-interest.

But alas, personalization happens in the digital world, too. In fact, some argue that digital communications may offer *more* intimate exchanges than their analog counterparts have offered, mentioning how the messages that are saved in smart-phones may be more intimate and explicit than those shared in bed (Perry 2017). Others (e.g., Beck and Beck-Gernsheim 2014) note that digital technology has allowed intimate partners split by geographic separation to be more intimate than in the past when only pen-and-paper letter writing was possible.

Maybe technology is "reshaping how we view romance" (Perry 2017). Even with digitization of love letters, the value of individualizing the exchange between two people is tremendously powerful. This technological reshaping and the cultural values at play are present in this observation by Perry (2017):

> Sure, internet interactions can come close to being the real thing, but what gets lost in translation from real world to the internet is physical human interaction. Social cues, facial expressions, actual touch, these are all communication signals that are just as important as words. These features of human interaction are difficult or even impossible to replicate over the internet. During an online exchange, it can easily be forgotten how our interactions are real and have real consequences. This can create a mind-set of treating online communication as a game, with the other person being turned into a game piece.

Part of personalization is the idea of connectedness—that someone is really present if their handwriting or other pieces of evidence that they touched the object are embodied in a love letter. This is why a Skype call feels more personal than a text—it more closely mimics face-to-face communication because people can see and/or hear bodily presence of a partner (Janning, Gao, and Snyder 2017). The quote from Perry above compares face-to-face communication and internet interactions. Interestingly, the same argument—that physical human interaction is better than any proxy communication where the person is not physically present—can be made against paper letter writing, too. In the end, this type of sentiment—that physical presence matters in romantic love—actually serves to reinforce the cultural value that romantic love must involve personalization more than it challenges the preferable format for demonstrating that value. Even if someone is not afraid of digitization, if they think it's important to include emojis and special font and content that includes information only the two people know about it, they also like personalization. To digitize love letters, then, does not necessarily do away with the value of individualization and personalization as a requirement in romantic love.

In both paper and digital love letters, the person is not actually physically present. Those who require pen-and-paper formats to meet the criterion of personalization may believe that seeing handwriting and other evidence that someone physically touched an object that is now in their hands makes it feel more person-like. Those who believe digital platforms have the capacity to be personalized may assert that an emoji performs the same function as scented stationery: it personalizes the love letter. But in neither case is the desire for personalization and individualization of the lover absent.

Pace

Some define a love letter based on how much time it takes to write, send, read, and respond; the slower the pace, the safer it is to call it a love letter. Someone may send a letter and then spend days or weeks awaiting a response, especially if they are geographically separated. They may even include mention of how long the response took in the content of the next letter. In terms of curatorial practices, maybe saved love letters are organized chronologically, with dates as markers of the rhythm of various relationship stages or gaps in time between correspondences.

If love letters are digitized, some may fear that the pace of romance would change from one of patience and associated thoughtfulness to one of speed and associated carelessness. The idea of a love letter as the result of hours or days of introspection, selective wording, and attention to depth while sitting alone at a writing table is challenged in the face of communication technology that allows for quick response, and sometimes even response while completing other tasks. Curatorial practices surrounding the saving and storing of love letters also require time and care, with quick dispossession or sloppy storage less desirable. Illouz (1997) writes that communication in romantic relationships "fulfills its instrumental function only to the extent that it also fulfills the expressive function of providing warmth and intimacy" (238), which aligns well with the idea of personalization mentioned above. If someone defines intimacy and warmth as requiring time, effort, and depth, they may be wary of saying that a text fulfills the function of a love letter. They may be afraid that a fast pace means less warmth. This fear is situated in a world that is increasingly perceived as hectic, overscheduled, rushed, and filled with a bombardment of messages on phones, TVs, and radio waves. In this line of thinking, the cultural value of taking time in a hectic digitized world in order to stay connected to people is threatened if the slow rhythm of love letter exchange goes away.

Those individuals who are less likely to cling to a pen-and-paper love letter as the only way to preserve the slow and thoughtful pace of an idealized romance may suggest that digitization of love letters does not *necessarily* disrupt this rhythm. With paper letters, the only choice is to wait days. With digital communication, that choice doesn't go away; it is offered along with other choices that are faster. As noted in Chapter 2 when I referenced Stanley's (2015) assertion about innovations such as the telegraph and postcards, change in communication speed has long concerned people, and people have adjusted each time. Of course, today, it is increasingly likely for people to expect faster responses. Nearly every aspect of people's lives today, whether or not it involves communication, includes the management of time in a rapidly changing (and rapid) world. Digital technology has indeed impacted the pace of people's lives. Digital communication is a part of this changing world, and perhaps makes a fast pace more possible. But it is not necessarily the sole cause of it. In terms of curatorial practices, for example, there is nothing intrinsic to digital storage that makes it more or less conducive to carefulness or slowness. Anyone who has spent hours trying to organize a computer folder of documents that span decades knows this.

Even with digitization of love letters, the value of demonstrating thoughtfulness that connotes taking time may still be more powerful than the impact of format, both in terms of the love letter exchange and in its curatorial practices. Evidence of this includes features in smartphone applications for couples to share love notes, photos, and other digital mementos that require time and thought to preserve and organize in order to reference them later on. The fear of losing carefulness, patience, and depth that is associated with the decline of pen-and-paper letter writing may be associated more with larger cultural shifts about time and pace than with specific formats of communication. Those who believe that digitized love letters should count as love letters may suggest that there is evidence of a desire for thoughtfulness and slowing down even in digital platforms. Yes, these communication formats are part of a change toward speed and efficiency, but they are not necessarily the change itself. Someone may conveniently ignore past moments when a change of pace was met with resistance and fear, and then was eventually absorbed into the milieu of "that's how things are." In doing so, that person may also conveniently ignore the notion that the proper use and measurement of time is collectively defined. As I discussed in Chapter 4, understanding time beyond an individual's conception of it in their own personal experience, then, is an important component of how (and when) culture works.

Preservation

The reason there have been countless studies of love letters, their writers, and their content about historical events when they were written is because scholars have access to letters that have been preserved over time, maybe within families, or maybe within museums. Stories that emerge from these letters have provided the basis for understanding the past, which affects how the present is interpreted.

As discussed with regard to the social construction of time and memory in Chapter 4, sometimes this interpretation is shaded by a nostalgic view that is more about what people would like to envision from yesteryear, rather than how it actually occurred. The preservation is about saving a story that matters now regardless of its accuracy, and saving the present story for an imagined future nostalgia among those who eventually are left behind.

Some may define a love letter as requiring the capacity to be preserved in order to make the love story outlive the lovers, either within a family who is invested in their ancestry, or in a larger group invested in its collective history. While it may seem unlikely for someone writing a love letter to think about their future children being able to know about the early stages of their parents' marriage by reading that letter (or about a future museum that may house their letter as an intriguing archeological symbol of the time period in which it is written), it is easier to imagine people saving a love letter as a way to see and access their own story. Sometimes this is about preserving a memory, which is what people do when they save souvenirs from travels. Sometimes this is about needing a material reminder of an identity that mattered at a certain time—a version of the self that has shaped who someone is today in their identity and in their relationships with others. And sometimes, even if unusually, when people spend time curating the letters so that they may be accessed by others, it is about a desire to have their story told beyond themselves in the future.

This is all really about having a legacy, which aligns with the element of personalization above, since having a legacy is about self-interest as much as it may be about the well-being of future generations (if that wasn't the case, there wouldn't be buildings named after wealthy individuals). People fear being insignificant, as not mattering. People aim for longevity because longevity is valued, and they fear losing their story if the story cannot be preserved using means that they understand or approve of.

But, some may ask, isn't preservation possible with digital formats, too? Don't people use digital means to preserve stories to be used to reminisce back in time, to think about today's stories, and to hope the story will continue into the future? Can't the electronic version of this book be shared with my future (hypothetical) grandchildren just as much as a hard copy could be? Either way, my hypothetical grandchildren have my book and know my story. What matters to people who do not adhere to the belief that an object needs to be in physical form in order to be preserved is not *how* someone's story will go on, but *whether* it will. In fact, for some, to preserve something digitally renders it even safer.

Someone may want to remember past stories because it helps them understand their own story today. From photo albums to school and travel mementos, people have both physical and digital reminders of where they come from. Importantly, some are selective with which reminders they seek and save, perhaps succumbing to the allure of nostalgia in thinking about what part of their past story they want to use in the present. Sometimes people want their lives to be remembered tomorrow, too, because having a legacy is valued in society. Curating this legacy enacts the cultural value that stories are meant to be preserved, and, often, that people only want the good stories about themselves to be part of that legacy. Perhaps the value of curating a legacy matters more than the platform that holds the stories. Expressing mistrust of the security of different platforms (paper can burn; electronic files can vanish) reveals the power of fear of losing stories and the value of security of important objects. This expression is more about the value of having a legacy than it is about the format that preserves it.

Privacy

To keep something private is to withhold sharing it with others who are defined as being excluded from an inner circle of some sort. If privacy matters to someone in their definition of love letter, they may view a couple's desire to keep their love letters hidden from others positively. In the writing or the storing (or, actually, even in the dispossessing) of a love letter, hiding it from people beyond the couple is a way to enact privacy.

Accidentally sharing intimate stories with outsiders when the stories have been defined as private may lead the possessor of the stories to fear outcomes such as embarrassment, the disclosure of damaging information, or the revelation of traits that they prefer to keep out of the public eye (e.g., sexually explicit messages found at work). To put it in terms of social theorist Erving Goffman's famed *dramaturgical* approach within sociology (real life is like a theatrical production, complete with roles, props, and stages), their back stage has become their front stage. The actor's nightmare has come true. If a child stumbles upon his parents' steamy love letters that contain reference to sexual behaviors (or if a boss finds these), and the family (or workplace) is one where discussions of sexuality are defined as private matters, then borders between stages have been disrupted. The belief in, and reinforcement of, the boundary between public and private (and the concomitant norm of treating romantic love as a private matter) is alive and well in contemporary Western society, even though individual relationships move, shape, and puncture that boundary in many ways. In essence, as soon as someone defines a part of their life as private, making pieces contained within it public can be painful and can violate norms (or even break rules) about what is supposed to be kept private.

Are digital love letters more or less private than paper ones? On one hand, "social media networks blur the boundaries and expectations of what should and should not be held private in a relationship" (Perry 2017), which may lend support to someone's argument if they fear the loss of love letters by virtue of them being digitized. More digitization may lead to more blurring of private and public lives. Certainly social media posts demonstrate this blurring well. On the other hand, it is a mistake to think that the exchanges of paper love letters were somehow always and ubiquitously private matters, with decades' old evidence of family members reading couples' romantic sentiments aloud, either to share other news contained within a letter, or because the recipient couldn't read (Stanley 2015). The monogamous dyad as the only people to ever write or read or store each other's love letters is an ahistorical ideal that erroneously assumes: (1) monogamy among all lovers; (2) literacy among all lovers; and (3) privacy from others beyond the lovers. Further, since affection is often displayed publicly and people are invested in that display (otherwise there would not be a term to describe it: *public display of affection*), it is clear that love is not always a private matter.

The fear of loss of privacy may have more to do with someone's definition of what counts as private than it does with the format that keeps it private. In this way, a person who wants to keep their love letters private while reading them at a coffee shop will prevent others from looking while they read. Whether the letters are read from a smartphone screen or from scented stationery, the reader will likely assume a posture and display gestures that signify to others the message of "Please don't look at what I'm reading." Hunching over a private love letter excludes public eyes from private words, but the love letter format may not affect the likelihood to hunch in the first place.

Place

Where love letters are located, especially in terms of curatorial practices once they've been received, gets at the significance of place. Someone may write a love letter in the hopes that the recipient places it with other keepsakes from the relationship; another person may keep all of the love letters from their high school sweethearts in one spot, while the ones from a spouse are kept in another. Maybe a collection of love letters is in a place that is hard to access without a lot of brain-wracking. Maybe after someone writes a love letter they worry about where their lover is going to store it, for fear (as connected with the notion of privacy) that someone will find it who shouldn't. Maybe someone displays a love letter prominently on their desk as a reminder of how cherished the person is who sent it. While this can look different from person to person, people singularize objects that are meaningful to them. A person will worry more about forgetting where a love letter is that has been defined as important (or from an important person) than they will about forgetting where their junk mail is (or where letters from a less significant person are). If people lose cherished objects, or put them in a precarious place where they get damaged or destroyed, they grieve their loss. Grief occurs, in part, because of the strong desire to have stories preserved. That's why people spend emotional labor thinking about how secure a love letter storage location is. And that's why people could answer survey questions about where their love letters are stored.

Is there a difference between worrying about the security of a physical versus a digital location? Which format affords more control over the place where treasured things are placed? Anyone who has ever lost an important computer file or fears that a confidential electronic document has been breached in a digital cloud knows how horrible the feeling of loss of control is. But is this more about the computer or cloud as the storage location or the fear of losing something important? Is the significance of place vastly different in digital versus physical locations? People worry about basement water damage ruining a box of letters; they worry about a digital cloud losing a file of letters. Perhaps, as discussed in Chapter 3 when I referenced the work of Belk (2013), who asserted that people can singularize both physical and digital objects, the worry is present in both cases even if the format differs. Identities and experiences may shape which format for preservation worries someone more (e.g., it is more likely for someone who has not grown up with digital file storage to mistrust digital security than it is for someone who only conducts banking online to have the same fear of loss). But the worry and the potential for the object to be made meaningful are still there in both formats. It is this worry that demonstrates how strong the cultural value of storing cherished possessions securely is.

***

While presented in sequence, it is important to note that these elements—personalization, pace, preservation, privacy, and place—influence and overlap with each other. For example, place (where love letters are stored) is impacted by cultural values about privacy, since storing possessions behind closed doors or in password-protected digital clouds can only happen if doors that close are valued and constructed in homes and only if clouds are valued and constructed for electronic files. One could imagine storing love letters from a particularly special person in a different location (place) from those written by people who are less special in order to heighten personalization of that particular person. The more invested someone may be in preservation of a love letter, the more attention they may give to finding a secure place to store it. If a love letter

appears to have been crafted with tremendous attention to detail, thus necessitating its creation to take more time than usual to make and send (pace), it
could be defined as more personalized by its recipient and therefore harder to
throw away or delete. If a person wishes to keep a love letter private but also
wants to preserve their story over time by putting it in a place where future family members can easily find them, they are demonstrating the interplay between
the definitional elements of privacy, place, and preservation.

In the initial paragraphs found just underneath each of the definitional elements presented above, there is no explicit mention of love letters represented
as either paper or digital formats. I did this on purpose so that different images
of love letters could appear to readers. Some may have envisioned only paper
love letters in the first few lines of each section above; some may have pictured
images of smartphone apps and words on computer screens. Which image
appears upon reading depends on how a reader's own culturally prescribed values about digital communications influence how they see the world (and what
they do or don't fear). The goal of this book is to show what may not be readily visible: how powerful culture is in seemingly individual experiences. How
fearful someone is about the digitization of love letters may depend on how
much they believe this may threaten larger cultural values that matter to them.
Pointing out *how culture works* as an exercise in self-understanding, rather than
saying that there is a right or wrong definition of love letter, is my central aim
in presenting this framework.

Despite my belief that the framework I present is a useful exercise in understanding how culture works, it would be irresponsible for me to assert wholeheartedly that format (digital or paper) makes *no* difference in terms of the
meaningfulness of love letters. In fact, it is important to recognize that the
increased digitization of love letters may impact the content of them so much
that the whole definition of love letter may change in years to come. Or it may
be that the possibility of calling a digital message a love letter at all will be
abandoned, as the writing, sending, and receiving of paper letters continues to
dwindle and be seen as rare and precious. Maybe the lessening of pen-and-paper love letter writing will preserve that format as the only thing that counts in
a definition of love letter, as if the definition will have been stopped in time as
its practice ended. My point here is that today marks an interesting crossroads
when it comes to the definition, with some people believing that a broader definition of love letter should include digital versions because of all of the reasons
contained in the framework I've presented. These people would argue that
digital love letters have the potential to: have the epistolary intent of being personalized and intimate; take any amount of time to exchange; be preserved; be
kept private; and be placed in a safe and important spot. But some people may

fear big shifts in cultural values that an inclusive definition would bring about, and wish to stick to the pen-and-paper love letter as the only kind that counts.

By studying love letter curatorial practices rather than content, this book adds to the ongoing discussion of the impact of technology in everyday life. What will be interesting to see over time is whether the ubiquity of digital communication use and the increase in numbers and types of digital platforms to exchange romantic messages will alter the content of love letters themselves, or even alter the cultural values that shape romantic love. As Hayles argued in the 2002 book *Writing Machines*, the meaning of words is impacted by the form (paper versus digital) those words take. However, if the proliferation of all of these new digital love letter platforms and the dwindling of pen-and-paper love letter writing doesn't make the definitional elements of love letters (personalization, pace, preservation, privacy, and place) disappear, then there will be evidence of the power of the cultural values about romantic love at play over the format that is used to convey these values.

Some people are trying to dispel some fear associated with the digitization of love letters. Margie Morris, a clinical psychologist and researcher at Intel trying to find ways for technology to "promote interpersonal connectedness," observed in a 2014 *Wired* online magazine article that "Love letters aren't over. They are just getting smarter, and more social." She claims this despite the exchange of billions of emails, text messages, and social media posts shared worldwide every day, and despite love letters being "reduced to simple two-line expressions that can never be read for specific, meaningful content." Her hope is that, in the face of the explosion of messages sent and the diminishment of lengthy content, technology can still help people read idiosyncratic communication cues using language style and emoticons. Perhaps, also, it can help people figure out ways to be smarter about flirting or expressing emotion in relationships that span the globe—relationships that are more prone to misinterpretation given the cultural influence of romantic expression. In the end, as the Intel project demonstrates, the cultural values that surround contemporary definitions of romantic love are more powerful than the digital or paper format.

To reiterate, my point is not to say that there is no impact of digital technology on everyday practices and relationships. It would be irresponsible to overstate that technology is somehow benign in its influence. My point, instead, is to say that sometimes a finger is pointed at the wrong thing. It's better to take a step back and think about what cultural values might matter in the estimation about the meaning of a particular object or fear of losing a particular meaning that someone may imbue in that object. Instead of saying that love letters are going away as a blanket statement, it is better to ask what elements are part of the definition of love letter in the first place. Maybe instead of asking whether saving a bunch of text messages is better or worse than saving a box of paper letters, the question should really be about what values are being upheld in the desire to save love letters at all.

The framework I present may be less about love letters and romance and more about larger Western values that shape a multitude of aspects of contemporary life. Think about the aforementioned values—self-interest, maintenance of an appropriate pace, preservation of one's story, privacy, and belief that objects need to be kept secure—as indicators of life in general. Might it be useful to broaden the lens to define love letters as important indicators of larger social patterns? Might it be useful to embed conceptions of romantic love in larger political, economic, and cultural systems that not only serve to shape everyday practices, but also shape the systems of inequality that give people more or less access to the things they love?

Privilege and the Fear of Love Letter Loss

By presenting a conceptual framework as an exercise in understanding how culture works, I demonstrate my own fear: I am afraid of the lack of understanding of cultural forces as people evaluate their lives. But I am even more afraid of the invisibility of privilege, which plays a prominent role in a lack of understanding of cultural forces. From the digital divide and global differences in literacy to the preservation of a heteronormative gendered household division of labor, I have discussed group patterns and inequalities that relate to love letters throughout this book. Privilege is access to rights or resource that are culturally valuable and that come based not on individual effort but on group membership. Understanding privilege shows how culture intersects with individual practices beyond just pointing out differences.

Differential access to rights and resources has implications for well-being. People may feel as if they are functioning well in a cultural context if they adhere to an accepted emotional display or experience, but not everyone has the same ability to adhere to what's accepted. Cultural values surrounding romantic love are so powerful that adherence to them is likely to enhance well-being because participation in a well-supported system can give someone a sense of belonging. Some may even say romantic love is functional, since it perpetuates human connectedness, legitimate means for creating families to reproduce society, and success in multiple industries that benefit from the buying and selling of goods and experiences that symbolize romantic love. People may believe this regardless of whether they prefer paper or digital love letters, as the previous section elaborated. However, in the claim that experiencing a culturally prescribed emotion is functional (to fit in, to accomplish a task, to accomplish well-being, to feel a sense of belonging or satisfaction), it is important to ask, "functional for *whom*?" Here I aim to make privilege visible in the framework of how fear of losing particular cultural values operates with the definition of love letters in terms of personalization, pace, preservation, privacy, and place in today's digital world. By doing this, I reaffirm the notion that romantic love has been defined as culturally valuable, I argue that groups have different access to upholding this value, and I offer ways that digitization may impact this differential access.

The definitional element of personalization of love letters includes the cultural values of individualization and self-interest. But, perhaps paradoxically, these values are supported in exchanges that are perceived as separate from a depersonalized marketplace at the same time they operate within it. Illouz's (1997, 2012) work on how social class impacts people's access to culturally accepted (and sometimes expensive) markers of romantic love (e.g., dates, travel, polished and educated communication styles), and countless scholars' writings on the historical changes in legal and social acceptance of romantic love among homosexual partners begs the question, who has access to the marketplace where romantic love is exchanged? Who has the right to operate within it? Who gets to experience and display elements of personalization to each other in ways that are deemed acceptable by others? Who may have greater access to the value of self-interest in their socialization process—a value that pays off better in a capitalist marketplace? If adhering to personalization is required in defining something as a love letter, then someone who is able to do this may gain a better sense of well-being than someone who is not, because they are adhering to this cultural value.

But, one may correctly assert, it is not as if it costs a lot of money to write a letter, fold it, put it in a stamped envelope, and mail it (it also doesn't cost a lot of money to send a text). What is at stake here is not just the affordability of market goods and services that demonstrate romance, though. Whether someone has access to the symbols and actions that constitute an accepted *expression* of romantic love is also at stake. Digital technologies are differentially available to different groups, to be sure. But their near ubiquitous presence and their capacity to offer the ability to exchange intimate communications can serve populations who may have been excluded from more mainstream displays of affection. In this way, it may actually be the case that defining love letters to include digital platforms allows for the democratization of personalization. However, the inclusion of a certain type of writing—formal, educated, following specific style conventions—in a definition of love letter may exclude people who do not have access to experiences or resources that make this type of communication happen.

Taking time is valued in a hectic digitized world, a value preserved in the fear of losing love letters by virtue of the sped up pace of their writing and sending.

But might time operate differently depending on a person's privilege? Some groups—especially affluent individuals, or individuals who do not perform shift work, for example—have greater access to being able to take time when they want it, or to having control over their own time. The digitization of love letters allows for quick exchanges that are accessible for a wide range of participants, whether it's lovers sending a quick message between work breaks or a wife sending her spouse a message through social media while she works in a different country caring for someone else's children. If a love letter definition needs to include taking time, it would exclude participation from people who don't have enough of it. The potential for digital communication to allow for quick exchange may lessen the divide between groups whose aim is to express romantic love to a partner.

In terms of the definitional element of preservation, having a legacy (and ensuring its longevity via documentation) is valued. But it is valued at the same time all past stories have been selectively curated over time. Think of all of the groups throughout history whose stories have been rendered silent or invisible because their writings either did not exist (because of unequal access to literacy or the right to write) or were not preserved (because their stories were not defined as valuable to preserve). The attention paid to selective parts of history and the allure of nostalgia create an ahistorical, mythologized, and idealized view of the past. This view—that only important things have been saved—serves to raise the privileged stories up as if they represented everyone's stories. The possibility of digital communication, and the ability to preserve mass amounts of information from vast amounts of people in very little physical space, allows for new and creative ways to capture the stories that were at one time kept hidden. Digitizing romantic communication, once saved, is not kept in a basement closet. And it may succumb to the same historical foibles of being lost or rendered unimportant by virtue of someone labeling them as such. But the capacity to share and save more messages from more kinds of people may make missing love stories less likely.

Maintaining the public-private boundary and the intimacy of the monogamous dyad is valued. But, not everyone has the same ability to preserve a boundary between their private and public lives. Privilege operates within the public-private boundary. On one hand, the display of romantic love among groups whose identities or relationships are deemed illegitimate in the eyes of their families or of the state, as has been prominent until recently with the legal classification of same-sex marriage or farther in the past with interracial marriages, can be relegated to the private realm. Keeping forbidden love hidden has been the marker of disenfranchised groups since people started defining what romantic love was. On the other hand, certain groups have been given less access to the means by which things can be kept private. Think about the visibility (and scrutiny) of private life that occurs while living in public housing, for example. Think about homeless individuals who have no private spaces to control. Think about the greater amount of privacy afforded to people who sit in the first class cabin in an airplane. In terms of love letters, then, someone's group status may actually impact whether and how much control they have over places, not only to store and access them personally, but also to protect them from the eyes of others whose judgments may otherwise serve to reinforce their disenfranchised status.

How do place and privilege intersect? Storage locations are shaped by values about architectural design, materialism, belief in the security of storage platforms, and methods of organization. These are embedded in privilege and inequality. Storing cherished possessions in secure locations is valued, and security is increasingly precarious. When someone does not have enough space, or control over a space, they are less able to control the saving and accessing of cherished possessions. Finding housing that is safely secured with door locks or having a job where internet and email security is paid for by the company are both privileges that not everyone can access. Thus, the curatorial practices surrounding the location of cherished objects, both physical and digital, are impacted by how privileged someone is in terms of access to, and control of, storage locations.

In Chapter 3 I discussed the cultural value of minimalism and the importance of home storage places in people's (especially women's) love letter curatorial practices. The overwhelming feeling that accompanies having too much stuff is present in many corners of contemporary culture. But to experience this feeling, of course, requires having the stuff in the first place and agreeing that it's a bad idea to have so much stuff. Having a home, enough space within a home to store non-instrumental objects such as love letters, and a collection of stuff that needs to be stored are all privileges. In a 2016 *New York Times* Op-Ed, Stephanie Land refers to the tidying up movement in terms of choice and social class. She notes, "minimalism is a virtue only when it's a choice, and it's telling that its fan base is clustered in the well-off middle class. For people who are not so well off, the idea of opting to have even less is not really an option." Thus, while control over stuff may bring emotional well-being, as decluttering experts may advise, the ability to control something requires having the stuff in the first place. These tidying projects need to be understood in light of the socioeconomic inequalities shaping the projects. Saving love letters in a secure place is not equally likely for all social groups. As the next decades unfold, it will be interesting to see whether the possibility of storing cherished possessions that tell the story of family and personal relationships digitally may allow for a wider variety of places that are accessible to a wider array of people across the socioeconomic spectrum. Perhaps also, the storage and organization of love letters as a gendered domestic practice may lessen as digital places become substitutes for physical places.

Conclusion

After I presented my research on love letters to an audience of older individuals, a man came up to me to share a story. He had been in the Vietnam War and had written pen-and-paper love letters frequently to the woman who became his wife. When he returned, he wondered where those letters had gone. His wife told him they were stored at her mother's house. As he continued telling the story, he leaned in to me across the table. This, he whispered, was mortifying. After all, what if his mother-in-law found them? They were filled with explicit sexual references about her daughter. His response to finding this out was to travel to his mother-in-law's house, grab the letters, and promptly burn them. To him, they were so private that he didn't want anyone to see them again. I still wonder what his wife thought of this.

Burning love letters to maintain their privacy came up recently, when a friend told me that she and her partner had burned their early-romance letters from the 1990s, but they did it together as a decisive act to solidify these as shared memories, but not shared beyond the two of them. It was the fear that their children may find them and read the steamy sexual exchanges that prompted this act.

In terms of the definitional elements discussed in the conceptual framework I propose in this chapter, what are these people afraid to lose? Certainly not the letters themselves, nor a desire to preserve a story for the next generation or for someone putting together a museum exhibit about the Vietnam War or early 1990s romances. They are also not afraid of losing connectedness with their partners. They are afraid of losing control of what they saw as a private collection. If given the option neither of these individuals would have chosen to digitize their love letter collections because to share them (or create the potential for wider sharing) would actually depersonalize them. For them, the best way to personalize these love letters was to get rid of them, rendering them so special that the only way to get to them was by their choosing to tell a story about them. By burning the letters, perhaps contrary to the cultural value of cherishing close-at-hand possessions that symbolize a relationship, these people actually adhered to the cultural value that sexual intimacy is still a private matter.

Few things feel as private as a romantic love note, but, as this book has hopefully shown, the elements of private lives tell a story that extends beyond the pages of one letter, or the boundaries of one romantic partnership. Norms and values associated with how romantic relationships operate in a digital world are still being sorted, and this sorting is not new. When postcards and the telegraph came along, people were worried that handwritten long-form letters would die out. What is new are the ever-changing array of formats, the workload and bodily tasks associated with handling and managing digital rather than physical stuff, and the mental energy trying to figure out if digitization matters in terms of how to define love letters. My task in this chapter has been to provide a framework for anyone who may be wondering where their mental energy is going in a concern about love letters fading away.

Importantly, the sorting of norms has to be situated in a particular time and place. What is considered proper letter etiquette in one place may not be defined as such in another. My goal for this book spans individual, interactional, and cultural locations. I hope that readers come away with greater self-awareness of romantic communication habits and preferences, including curatorial practices. I hope that readers see that there is still choice present in terms of love letters, especially in terms of how to interpret fears, manage expectations in light of cultural influences, recognize how privilege may operate in everyday life in hidden ways, and communicate with others about all of these. In other words, choice is in how culture is used once it is understood. This quote from blogger Leah Perry (2017) captures my goal well:

> Despite the obvious ways in which romance has evolved over time, it is also evident that the way we function as a society has evolved with the use of technological devices. With the negative ways that technology has impacted romance taken into consideration, it should also be noted that technology has provided effective ways in creating new and meaningful relationships. From my analysis, I leave with one observation: maybe romance has not been ruined, but redefined.

For students of sociology and anthropology, this book sheds light on important steps needed in order to approach the world with a keen *sociological imagination*, or the ability to understand personal stories in light of a larger sociohistorical context (Mills 1959): first, to recognize that emotions are experienced individually but also shape and are shaped by groups and collectives; second, to understand how individual experience of emotion shows culturally prescribed values that affect the well-being of individuals; and third, to recognize that well-being is a privilege that is differentially accessed and experienced by people in different groups. Who knew that something as seemingly small and private as a love letter could say so much about the social world?

For people interested in how this book may assist in their own romantic pursuits or self-exploration about romantic goals, my advice is to work on defining personal preferences and be able to communicate these to a partner in order to bridge the gap between ideal and reality. If the people taking my survey are any indication of this gap, there is a chasm between what people want (decorated handwritten paper love letters) and what people actually do (send a quick text to say *I love you* and *please pick up some rice at the store*). People should be able to define their fears in terms of personalization, pace, preservation, privacy, or place. The best remedy for allaying fear is to discover what is really happening. Hopefully this book has offered at least some new part of the story of that reality and a framework to understand it, so that people can operate with a bit more information in their own investigations into romantic love.

Usually, near the conclusion of a piece of research, an author will offer a long list of limitations to address what their research did not do and to offer readers ideas for future research of questions that have yet to be answered. My list of limitations, admittedly, is quite long. I did not do ethnographic research, which disallows the possibility of thick and rich description of how culture works. I did not include every type of love letter platform that exists, in part because the technology (not to mention the name of applications) changes so rapidly. I did not study cognitive or neurological processes that capture how brains interpret pen-and-paper versus digital messages, nor how brains may be impacted differently depending on curatorial practices surrounding physical versus digital objects. I did not conduct new research outside of the United States, rendering claims about Western ideals of romantic love both appropriate and terribly limiting in light of increased global connectedness (and inequality) in today's digital world. I did not ask all of the questions I could have in my survey, including more open-ended questions about why people get rid of their love letters (rather than what they do with saved ones).

The people whose stories are in these pages, both from my survey respondents and from those whose paths I crossed in my own life, represent a close link to the demographic groups that I occupy. Such is the nature of convenience sampling and the drawbacks of researchers relying, perhaps too much, on their own social networks to answer their research questions. Hopefully by including other scholars' research, as well as their justification of the presentation of data from similar groups, this limitation is at least contextualized.

I didn't differentiate type or length of romantic relationships in my investigation about love letter curatorial practices. Future research could examine this, since the expression of love must also be situated in norms and rules about how that love may be expressed. For example, it is different for an unmarried person to send (or save) a love letter to a lover at a time when marriage is expected than it is to send one in the midst of increasing diversity of romantic partnerships that are seen as acceptable. The same can be said for the level of sexual explicitness in letters. References to sex have been in love letters for a long time, but the likelihood that someone may be embarrassed if someone finds the letters depends on how embarrassing those references are. Acceptance levels and definitions of sexual explicitness have changed over time and across geographic space (as well as between different groups), suggesting that cultural context may shape this element of love letter content, which may in turn affect curatorial practices. Belief that the relationship or its level of sexual explicitness referenced in a love letter is acceptable may even impact the likelihood of someone sharing their love letters with others as opposed to keeping them entirely private. Further, dating partners who have been together only a few months likely have different communication and saving practices than those who've been together for twenty years. While my central aim is to point out the overarching ways that romantic love operates in contemporary culture, I acknowledge wholeheartedly that, as Stephanie Coontz (2006) has shown, love has "conquered marriage." Clearly there can be structural reasons that impact the practices surrounding love letters.

Finally, I didn't address much in terms trauma or sexual violence via acts of sexual aggression, bullying, or harassment that can occur under the guise of "romantic communication," whether it be saved sexting photos used in revenge porn, online shaming or harassment of a former partner in public places such as Facebook, or the use of written communication in damaging ways in legal disputes involving partner loss, estrangement, or abuse. This limitation, thus, renders my work in jeopardy of romanticizing romantic love, which runs counter to my aims. While I acknowledge my tendency to focus away from relationship woes, doing so allows for the isolation of ideas that form the basis of the conceptual framework I apply. This, I hope, can serve as a starting point for others interested in related topics to test its application beyond the heart-shaped boxes under the bed.

People love love—the idea of it, the experience of it, the stories about it, and the symbols that represent it. It's hard to take a step back and deconstruct something that is so valued and valuable. Sometimes, when I present ideas such as those contained in this book, I am accused of taking the wind out of the sails of romantic love. After all, to suggest that love letters are mere conduits of the sometimes problematic cultural values associated with romantic love is to lessen their specialness. Or could it be the opposite? Couldn't a greater understanding of how culture works render them even more special? To raise love letters to the level of importance in political, economic, technological, and familial realms is precisely the thing that will raise their importance. So, of course there's room for romance for social researchers! But maybe the more intriguing romance lies in the appreciation for how people can understand themselves and how they show their humanity to each other. Isn't it romantic to understand how people operate, and to understand the self in the process? That, perhaps, is the best way to gain depth in contemporary relationships. My wish for anyone who reads this book is to fall in love with posing hard questions that make people get to know themselves and each other a little better. It makes my heart sing to know that something as small as a love letter can tell the big story of how culture works.

Notes

1 The book *Love Letters: An Anthology of Passion* (Lovric 1995) captures the content of love letters. Lovric organizes love letters from famous historical men by their epistolary content, including those that express admiration, proposition, possession, frustration, and even confrontation. While my research does not capture epistolary content other than perhaps indirectly in cases where people revisit letters in order to remind themselves what to avoid in future relationships (or in stories mentioned earlier when people wanted to get rid of love letters because of their sexual content), it would be interesting to examine whether the epistolary content shapes the likelihood to save a love letter, as well as the curatorial practices once saved. In other words, future research could connect epistolary intent more closely with curatorial intent by asking people whether there was something about the content that made them more likely to save a love letter, reread it, organize it in a certain way, or even burn it.

2 This framework emerged inductively as a result of the iterative and multi-month process of data interpretation, writing, and consulting with others' research. To check the internal consistency and sensibility of the framework's application to love letters and other topics, I held a "teaching session" with nine undergraduate sociology majors whereby I presented the central themes and findings of the book and explained the conceptual framework. I then incorporated student feedback into an improved rendition. Thus, while this framework emerged from empirical research findings, it has not been tested empirically with new data. However, the conceptual framework has been tested "in the field" to see if it: (1) applies to other topics, thus heightening its explanatory potential; and (2) makes sense both conceptually and pedagogically to undergraduate students, who are an intended audience for this book.

Methodological Appendix

SURVEY DESIGN, DEMOGRAPHICS, AND DATA ANALYTIC TECHNIQUES

Operationalizing Romantic Love and Love Letters

Throughout this book I present findings from a survey project that I conducted in 2013. In the survey, I provided a description of love letters without using that actual term, precisely because I didn't want people to exclude some forms of communication that would result from a narrow definition of love letter. In fact, the title of the survey as presented to potential respondents was "Communication in Relationships." For this reason, and yet to be sure the thematic content that came to mind for survey respondents was at least somewhat uniform in terms of connection to romantic love (their own experience with it or a hypothetical one), I defined the types of communication I wanted respondents to consider by introducing the topic in this way:

> The next questions are about different ways people communicate when they are in a relationship . . . communication between romantic partners that is NOT face-to-face communication, and . . . communication that is NOT JUST about logistics, scheduling, reminders, etc.

This means that messages exchanged between romantic partners that contain grocery lists, content unrelated to romance, and/or reminders can count in this definition only if at least some part of the message pertains to romance, intimacy, or affection.

In order to ask about love letters (without even mentioning that term), I also needed to offer survey respondents a clear definition of romantic relationship, since that, too, is a highly subjective term. As Teo (2005) offers,

> The term "romantic love" has almost as many definitions as there are authors writing on the subject, but I use it loosely to refer to the wide range of expressions and practices of affectionate attraction between . . . people, and to distinguish it from other types of love such as love between family members or friends.

(343)

In my survey, after defining the communication itself, I proceeded to ask people who had been involved in at least one romantic relationship to "think back on your relationship history and *select one relationship (previous or current)* to use for the remainder of this survey. Please try to choose a relationship where you can comment on *communication between partners* as described above." They could answer questions based on a past or present relationship. This question was prefaced with the following guiding words, which align with Teo's (2005) definition: "It's hard to provide a definition of 'Romantic Relationship.' For this survey, [it is defined] to include physical intimacy, attraction, commitment, and development over time." Acknowledging that relationship length and level of seriousness and depth are variable from person to person, I asked respondents to consider relationships that had lasted at least three months because I was interested in studying communication *patterns*, something that is harder to discern in a short amount of time.

After these clarifying introductory comments, I asked questions about curatorial practices surrounding the communications from a romantic relationships. I started by asking "Have you saved any communications from this relationship?" The next questions then asked about the saving, storing, sharing, re-reading, organizing, and other practices related to the saved communication, including letters, cards, handwritten notes, emails, text conversations, captured Snapchats, and Facebook message conversations.

While at first glance these questions may suggest that "anything goes" when it comes to love letters in my study, responses that occurred later in the survey about their practices surrounding saved communications (and other saved mementos) suggest that, while "love letter" was not explicitly mentioned, the content of the letters connoted romantic love per the aforementioned definition. These responses make up the bulk of the research findings presented in this book. I present findings that are both quantitative (based on numeric analyses) and qualitative (based on responses to open-ended questions that allowed survey respondents to elaborate and share their stories and ideas in narrative form; these responses were organized into themes and patterns, and the clusters of similar responses were also used in numeric analyses).

A portion of the survey not discussed in this book is a section with questions about long-distance relationships and relationships with long-distance segments, which is the subject of Janning et al. (2017). I exclude these findings in this book because they did not assess curatorial practices surrounding love letters as objects; rather, they got at routines, practices, and beliefs about the communication process itself. More specifically, I researched what communication formats people used when they were apart and how preferred formats were more or less meaningful to them. This included telephone and video/webcam calls along with written communication.

Importantly, conducting a survey and reporting quantitative results for a large sample of people is common in sociology, and less so in anthropology, although the methods used by scholars in both disciplines overlap in many ways. If an anthropologist wrote this book, likely it would include much more detailed description and interpretations of qualitative data, probably collected in in-person interviews and ethnographic observations. The benefit of a survey is that I could capture a lot of people's responses—several hundred, in fact—in a short amount of time. The drawback, which would be remedied with a more qualitative ethnographic project, is that I did not get in-depth responses, nor was I able to follow up with respondents to ask them more details about their stories. I did not get to meet respondents in person to talk with them about their subjective experiences with love letters. Because of these drawbacks, I have supplemented my survey findings throughout the book with news stories and anecdotes from people I've met while talking about my research. I interpret all of this in light of past scholars' theories and research findings. I use all of this to formulate a conceptual framework to understand how love letters demonstrate how culture works in light of rapid changes in digital communication technology.

Survey Demographics

During July 2013 I distributed an online Qualtrics survey about romantic communications to people via a college student listserv and via my and my research assistant's social networking sites. Once the survey had been open for a few days, in order to yield a more diverse sample in terms of age, race, geography, and gender, I sent it to strategically identified groups who were underrepresented in the original survey responses. In social scientific methodological terms, this made the sample purposive and nonrandom. My survey responses are from a convenience or snowball sample.

The data for this analysis come from 373 U.S. adults who have been involved in at least one romantic relationship where digital and/or paper communications from a partner were saved (for Chapter 4, the findings are based on the responses of 487 people, not all of whom answered the questions that are part of other analyses, which is why for most analyses the sample size is 373). Respondents are between the ages of 18 and 86, with just under 50 percent between the ages of 18 and 25, and the remaining half distributed evenly along each age category. For 64 percent of the respondents, the relationship they referenced in the survey started since 2005, with the remaining relationship start dates evenly distributed throughout the years. The earliest relationship start date referenced was 1954. When asked whether they do live or did live with this person at any point in the relationship, 53 percent said *yes* and 47 percent said *no*. Nearly 42 percent of the respondents have been in one or two romantic relationships in their lifetimes, 40 percent have been involved in three, four, or five relationships, and 17 percent have been involved in in six or more. The current relationship status (not necessarily referring to the relationship referenced in the questions about saving practices) for respondents is as follows: 23 percent are single, 37 percent are married, 39 percent are in a romantic relationship lasting more than three months, 4 percent are separated or divorced, 2 percent are remarried, and 1 percent are widowed. The remainder did not answer this question or listed something else.

The survey respondents are mostly women (79 percent), with 19 percent identifying as men, and the remainder choosing not to respond or listing something else. The racial-ethnic identity most represented is White (92 percent), with 6 percent Asian-American, 3 percent Hispanic or Latino/a, 2 percent American Indian or Alaska Native, 1 percent Black or African American, 1 percent Native Hawaiian or Other Pacific Islander, and 2 percent choosing something else (people could select all that apply, making the percentages add up to more than 100 percent). Just under 39 percent have attended some college, 28 percent have a bachelor's or associate degree, 27 percent have a graduate degree, and 4 percent have a high school degree, with the rest choosing other or not answering. In terms of annual household income, 13 percent list under $50,000; 28 percent list $50,000–$99,999; and 45 percent list $100,000 or higher, with the remaining 14 percent leaving this question blank. While I did not explicitly ask about sexual orientation, I asked people to identify the gender of their romantic partner. In terms of the romantic relationship that respondents had in mind for the questions about curatorial practices, 6 percent are men who reference male partners and 4 percent are women who reference female partners.

Only respondents who have saved any digital or paper communications from a salient romantic relationship are included in the findings in this book. To get at context in terms of general object saving practices, respondents were also asked these two questions: "Have you saved any physical or digital mementos that you would define as symbols of the relationship?" And "How would you describe yourself? I keep objects that are meaningful to me" (with a five-point scale of choices that contained the phrases "This definitely describes me" to "This definitely does NOT describe me" on either end).

Data Analytic Techniques

For all closed-ended measures discussed above, I calculated frequencies. In categories where there are very few people (e.g., in the "other" category for gender identification), I omitted them from analyses that separate them as a group. This is to protect their identity and preserve confidentiality, a practice common in the reporting of survey findings. In some cases, I collapsed response choices into broader categories (e.g., in some analyses specific communication formats are collapsed into the broad format categories of paper [letters, notes, cards] and digital [emails, texts, Facebook messages, captured Snapchats]). The qualitative analysis of open-ended questions consisted of two rounds of pen-and-paper open coding (Berg and Lune 2011) for thematic clusters of responses and a subsequent check for consistency between coders (me and a research assistant). To do this, I used the framework for qualitative data analysis developed by Miles, Huberman, and Saldaña (2013) to collapse respondent-level narrative data into coded aggregate themes, which then allowed for subsequent coding to note greater or lesser frequency of certain themes over others.

It is important to note that this study's non-probabilistic sampling method and lack of demographic representativeness means that generalizability cannot be assumed. In other words, this is a sample that may not represent the U.S. population more broadly. Additionally, given the speed with which new ICT are introduced into the marketplace, there are communication formats that have emerged even since the data for this project was collected in 2013 that are not included in this analysis. The data analyzed in this study also does not account for variation in communication formats due to differential access (e.g., some digital formats are more difficult to access in different geographic locations). I elaborate these and other limitations in Chapter 5.

BIBLIOGRAPHY

Ahearn, Laura M. 2001. *Invitations to Love: Literacy, Love Letters, and Social Change in Nepal.* Ann Arbor: University of Michigan Press.

Anderson, Monica, and Andrew Perrin. 2016. "13% of Americans Don't Use the Internet: Who Are They?" *Pew Research Center* (September 7). Retrieved January 31, 2018, www.pewresearch.org/fact-tank/2016/09/07/some-americans-dont-use-the-internet-who-are-they/.

Ansari, Aziz, with Eric Klinenberg. 2015. *Modern Romance.* New York: Penguin Books.

Appadurai, Arjun. 1986. "Introduction: Commodities and the Politics of Value." Pp. 3–63 in *The Social Life of Things*, edited by A. Appadurai. Cambridge: Cambridge University Press.

BBC News. 2013. "Languages of Love: 10 Unusual Terms of Endearment." (May 30). Retrieved January 23, 2018, www.bbc.com/news/magazine-22699938.

Beck, Ulrich, and Elisabeth Beck-Gernsheim. 2014. *Distant Love: Personal Life in the Global Age.* Translated by Rodney Livingstone. Cambridge: Polity Press.

Belk, Russell W. 2013. "Extended Self in a Digital World." *Journal of Consumer Research* 40: 477–500.

Bennett, Tony. 2007. "The Work of Culture." *Cultural Sociology* 1: 31–47.

Berg, Bruce L., and Howard Lune. 2011. *Qualitative Research Methods for the Social Sciences*, 8th edition. Upper Saddle River, NJ: Pearson.

boyd, danah. 2014. *It's Complicated: The Social Lives of Networked Teens.* New Haven and London: Yale University Press.

Brandt, Clare. 2006. "Devouring Time Finds Paper Toughish: What's Happened to Handwritten Letters in the Twenty-First Century?" *Auto/Biography Studies* 21(1): 7–19.

Bytegeist Podcast. 2017. "Talking Love Letters in the Digital Age with AMNH's Iris Lee." Retrieved January 28, 2018, https://soundcloud.com/librarybytegeist/5-talking-love-letters-in-the-digital-age-with-amnhs-iris-lee.

Carroll, Katherine. 2015. "Representing Ethnographic Data Through the Epistolary Form: A Correspondence Between a Breastmilk Donor and Recipient." *Qualitative Inquiry* 21(8): 686–695.

Coffin, Judith G. 2010. "Sex, Love, and Letters: Writing Simone de Beauvoir, 1949–1963." *The American Historical Review* 115(4): 1061–1088.

Collins, Lauren. 2013. "The Love App: Romance in the World's Most Wired City." *The New Yorker*, November 25, pp. 88–95.

Coontz, Stephanie. 2006. *Marriage, a History: How Love Conquered Marriage.* New York: Penguin Books.

Coontz, Stephanie. 2016. *The Way We Never Were: American Families and the Nostalgia Trap*, 2nd edition. New York: Basic Books.

Denegri-Knott, Janice, Rebecca Watkins, and Joseph Wood. 2012. "Transforming Digital Virtual Goods Into Meaningful Possessions." Pp. 76–91 in *Digital Virtual Consumption*, edited by M. Molesworth and J. Denegri-Knott. London: Routledge.

du Plessis, Erik Mygind, and Pelle Korsbæk Sørensen. 2017. "An Interview With Arlie Russell Hochschild: Critique and the Sociology of Emotions: Fear, Neoliberalism and the Acid Rainproof Fish." *Theory, Culture & Society* 34(7–8): 181–187.

Emery, Léa Rose. 2017. "What Modern Arranged Marriages Really Look Like." *Brides* (September 9). Retrieved January 17, 2018, www.brides.com/story/modern-arranged-marriages.

Epp, Amber M., and Linda L. Price. 2010. "The Storied Life of Singularized Objects: Forces of Agency and Network Transformation." *Journal of Consumer Research* 36(5): 820–837.

Ferudi, Frank. 1997. *Culture of Fear: Risk-Taking and the Morality of Low Expectations.* London and Washington: Cassell.

Fox, Susannah, and Lee Rainie. 2014. "The Web at 25 in the U.S." *Washington, D.C.: Pew Research Center.* Retrieved May 18, 2016, www.pewinternet.org/2014/02/27/the-web-at-25-in-the-u-s/.

Garfield, Simon. 2013. *To the Letter: A Journey Through a Vanishing World.* Edinburgh: Canongate.

Glassner, Barry. 1999. *The Culture of Fear: Why Americans Are Afraid of the Wrong Things.* New York: Basic Books.

Guglielmetti, Petra. 2018. "Project Declutter." *Real Simple* (January): 86–95.

Gullestad, Marianne. 2004. "Imagined Childhoods: Modernity, Self and Childhood in Autobiographical Accounts." Institute for Social Research.

Haggis, Jane, and Mary Holmes. 2011. "Epistles to Emails: Letters, Relationship Building and the Virtual Age." *Life Writing* 8(2): 169–185.

Hayles, N. Katherine. 2002. *Writing Machines.* Boston, MA: MIT Press.

Hepper, Erica G., Timothy D. Ritchie, Constantine Sedikides, and Tim Wildschut. 2012. "Odyssey's End: Lay Conceptions of Nostalgia Reflect its Original Homeric Meaning." *Emotion,* 12(1): 102–119.

Hochschild, Arlie Russell. 2003. *The Commercialization of Intimate Life: Notes From Home and Work.* Berkeley: University of California Press.

Hudspeth, Christopher. 2017. "These Three Questions Will Tell You How Long to Wait Before Responding to the Person Who Just Texted You." *BuzzFeed* (January 18). Retrieved January 23, 2018, www.buzzfeed.com/christopherhudspeth/how-long-should-you-wait-to-text-them-back?utm_term=.aiMNNWX3P5#.bcn226Qmgy.

Illouz, Eva. 1997. *Consuming the Romantic Utopia: Love and the Cultural Contradictions of Capitalism.* Berkeley: University of California Press.

Illouz, Eva. 2012. *Why Love Hurts: A Sociological Explanation.* Cambridge: Polity.

Janning, Michelle. 2009. "The Efficacy of Symbolic Work-Family Integration for Married Professionals Who Share Paid Work." *Journal of Humanities and Social Sciences* 3(1).

Janning, Michelle. 2015. "An Unexpected Box of Love Research." *Contexts* 14(1): 76.

Janning, Michelle. 2017. *The Stuff of Family Life: How Our Homes Reflect Our Lives.* Lanham, MD: Rowman & Littlefield.

Janning, Michelle, and Neal Christopherson. 2015. "Love Letters Lost? Gender and the Preservation of Digital and Paper Communication from Romantic Relationships." Pp. 245–266 in *Family Communication in an Age of Digital and Social Media,* edited by Carol J. Bruess. New York: Peter Lang International.

Janning, Michelle, Caitlyn Collins, and Jacqueline Kamm. 2011. "Gender, Space and Material Culture in Divorced Families." *Michigan Family Review* 15(1): 35–58.

Janning, Michelle, Wenjun Gao, and Emma Snyder. 2017. "Love Letters in the Digital Age: Meaningfulness in Long-Distance Romantic Relationship Communication Formats." *Journal of Family Issues.* Retrieved January 28, 2018, https://doi.org/10.1177/0192513X17698726.

Janning, Michelle, and Helen Scalise. 2015. "Gender and Intensive Mothering in Home Curation of Family Photography." *Journal of Family Issues* 36(12): 1702–1725.

Janning, Michelle, and Maya Volk. 2017. "Where the Heart Is: Home Space Transitions for Residential College Students." *Children's Geographies* 15(4): 478–490.

Jenkins, Carrie. 2017. *What Love Is and What It Could Be.* New York: Basic Books.

Jolly, Margaretta. 2011. "Lamenting the Letter and the Truth About Email." *Life Writing* 8: 153–167.

Kondo, Marie. 2014. *The Life-Changing Magic of Tidying Up: The Japanese Art of Decluttering and Organizing.* Berkeley, CA: Ten Speed Press.

Kopytoff, Igor. 1986. "The Cultural Biography of Things: Commoditization as Process." Pp. 64–94 in *The Social Life of Things*, edited by Arjun Appadurai. Cambridge: Cambridge University Press.

Land, Stephanie. 2016. "The Class Politics of Decluttering." *New York Times* (July 18). Retrieved January 24, 2018, www.nytimes.com/2016/07/18/opinion/the-class-politics-of-decluttering.html.

Licoppe, Christian. 2004. "'Connected' Presence: The Emergence of a New Repertoire for Managing Social Relationships in a Changing Communication Landscape." *Environment and Planning D: Society and Space* 22: 135–156.

Lovric, Michelle. 1995. *Love Letters: An Anthology of Passion.* Llanwellyn, PA: Shooting Star Publishing.

Lyons, Martyn. 1999. "Love Letters and Writing Practices: On *Écritures Intimes* in the Nineteenth Century." *Journal of Family History* 24(2): 232–239.

McMullen, John F. 2016. "The Digital Divide: A Technological Generation Gap." *Techopedia* (September 30). Retrieved January 24, 2018, www.techopedia.com/the-digital-divide-a-technological-generation-gap/2/29295.

Mesquita, Batja, Michael Boiger, and Jozefien De Leersnyder. 2016. "The Cultural Construction of Emotions." *Current Opinion in Psychology* 8: 31–36.

Miles, Matthew. B., A. Michael Huberman, and Johnny Saldaña. 2013. *Qualitative Data Analysis: A Methods Sourcebook*, 3rd edition. Thousand Oaks, CA: Sage Publications.

Miller, Daniel. 1997. *Material Culture and Mass Consumption.* New York: Blackwell.

Mills, C. Wright. 1959. *The Sociological Imagination.* Oxford: Oxford University Press.

Milner, Murray. 2016. *Freaks, Geeks, and Cool Kids: Teenagers in an Era of Consumerism, Standardized Tests, and Social Media.* New York: Routledge.

Mitchell, Amy, Katie Simmons, Katerina Eva Matsa, and Laura Silver. 2018. "Across Countries, Large Demographic Divides in How Often People Use the Internet and Social Media for News." *Pew Research Center* (January 11). Retrieved January 18, 2018, www.pewglobal.org/2018/01/11/detailed-tables-global-media-habits/.

Morris, Margie. 2014. "Love Letters Aren't Over—They're Just Smarter, More Social." *Wired* (February). Retrieved January 24, 2018, www.wired.com/insights/2014/02/digital-language-love/.

Nelson, Kris. 2015. "What Is Heteronormativity—And How Does It Apply to Your Feminism? Here Are 4 Examples." *Everyday Feminism* (July 24). Retrieved January 23, 2018, https://everydayfeminism.com/2015/07/what-is-heteronormativity/.

Perry, Leah. 2017. "From Love Letters to Facebook Messages: Has Technology Ruined Romance?" *Hastac.org* (November 30). Retrieved January 17, 2018, www.hastac.org/blogs/leahperry95/2017/11/30/love-letters-facebook-messages-has-technology-ruined-romance.

Petrelli, Daniella, and Steve Whittaker. 2010. "Family Memories in the Home: Contrasting Physical and Digital Mementos." *Personal Ubiquitous Computing*, 14(2): 153–169.

Rainie, Lee, and Andrew Perrin. 2017. "10 Facts About Smartphones as the iPhone Turns 10." *Pew Research Center* (June 28). Retrieved January 18, 2018, www.pewresearch.org/fact-tank/2017/06/28/10-facts-about-smartphones/.

Roman, Laura. 2018. "How Apps Helped Log One Long-Distance Couple's 'Love Letters of Our Time'." *National Public Radio* (January 21). Retrieved January 22, 2018, www.npr.org/sections/alltechconsidered/2018/01/21/579086927/how-apps-helped-log-one-long-distance-couples-love-letters-of-our-time.

Rutter, Virginia, and Pepper Schwartz. 2011. *The Gender of Sexuality: Exploring Sexual Possibilities*, 2nd edition. Lanham, MD: Rowman & Littlefield.

Shoaff, Morgan. 2015. "The Way This Husband Is Honoring His Late Wife Is a Beautiful Testament to the Power of Love." *Upworthy* (November 25). Retrieved January 22, 2018, www.upworthy.com/the-way-this-husband-is-honoring-his-late-wife-is-a-beautiful-testament-to-the-power-of-love?c=ufb3.

Siddiqui, Shakeel, and Darach Turley. 2006. "Extending the Self in a Digital World." Pp. 647–648 in *Advances in Consumer Research, Volume 33*, edited by C. Pechmann and L. Price. Duluth, MN: Association for Consumer Research.

Smith, Aaron. 2015. "U.S. Smartphone Use in 2015." *Washington, D.C.: Pew Research Center*. Retrieved May 18, 2016, www.pewinternet.org/2015/04/01/us-smartphone-use-in-2015/.

Smith, Aaron, and Monica Anderson. 2016. "5 Facts About Online Dating." *Pew Research Center* (February 29). Retrieved January 18, 2018, www.pewresearch.org/fact-tank/2016/02/29/5-facts-about-online-dating/.

Spillman, Lynette. 2001. *Cultural Sociology*. New York: Routledge.

Stanley, Liz. 2015. "The Death of the Letter? Epistolary Intent, Letterness, and the Many Ends of Letter-Writing." *Cultural Sociology* 9(2): 240–255.

Swidler, Ann. 2001. *Talk of Love: How Culture Matters*. Chicago: University of Chicago Press.

Tamboukou, Maria. 2011. "Interfaces in Narrative Research: Letters as Technologies of the Self and as Traces of Social Forces." *Qualitative Research* 11(5): 625–641.

Teo, Hsu-Ming. 2005. "Love Writes: Gender and Romantic Love in Australian Love Letters, 1860–1960." *Australian Feminist Studies* 20(48): 343–361.

Tudor, Andrew. 2003. "A (Macro) Sociology of Fear?" *The Sociological Review* 51(2): 238–256.

Turkle, Sherry. 2011. *Alone Together: Why We Expect More From Technology and Less From Each Other*. New York: Basic Books.

Voo, Jocelyn. 2008. "Arranged Marriage Gets High-Tech Twist." *CNN.com* (April 23). Retrieved January 17, 2018, www.cnn.com/2008/LIVING/personal/04/23/web.arranged.marriages/index.html.

Weisberg, Mitchell. 2011. "Student Attitudes and Behaviors Towards Digital Textbooks." *Publishing Research Quarterly* 27(2): 188–196.

Zaltzman, Helen. 2017a. "Open Me Part I." *The Allusionist Podcast* (October 27). Retrieved January 8, 2018, www.theallusionist.org/allusionist/open-me-1.

Zaltzman, Helen. 2017b. "Open Me Part II." *The Allusionist Podcast* (October 27). Retrieved January 8, 2018, www.theallusionist.org/allusionist/open-me-2.

Zillian, Nicole, and Eszter Hargittai. 2009. "Digital Distinction: Status-Specific Types of Internet Usage." *Social Science Quarterly* 90(2): 274–291.

INDEX

Page numbers in italics indicate figures on the corresponding pages.